Cambridge
checkpoint

W0036205

Lower Secondary
Science

8

THIRD EDITION

Lower Secondary
Science
8

Peter D Riley

HODDER
EDUCATION
AN HACHETTE UK COMPANY

To Tabitha, Holly and Pippa

Cambridge International copyright material in this publication is reproduced under licence and remains the intellectual property of Cambridge Assessment International Education.

The Boost knowledge tests and answers have been written by the authors. These may not fully reflect the approach of Cambridge Assessment International Education.

Every effort has been made to trace all copyright holders, but if any have been inadvertently overlooked, the Publishers will be pleased to make the necessary arrangements at the first opportunity.

Although every effort has been made to ensure that website addresses are correct at time of going to press, Hodder Education cannot be held responsible for the content of any website mentioned in this book. It is sometimes possible to find a relocated web page by typing in the address of the home page for a website in the URL window of your browser.

Third-party websites and resources referred to in this publication have not been endorsed by Cambridge Assessment International Education.

Hachette UK's policy is to use papers that are natural, renewable and recyclable products, made from wood grown in well-managed forests and other controlled sources. The logging and manufacturing processes are expected to conform to the environmental regulations of the country of origin.

Orders: please contact Hachette UK Distribution, Hely Hutchinson Centre, Milton Road, Didcot, Oxfordshire, OX11 7HH. Telephone: +44 (0)1235 827827. Email education@hachette.co.uk Lines are open from 9 a.m. to 5 p.m., Monday to Friday. You can also order through our website: www.hoddereducation.com

ISBN: 978 1 3983 0209 9

© Peter D Riley Ltd 2022

First published in 2005
Second edition published in 2011
This edition published in 2022 by
Hodder Education,
An Hachette UK Company
Carmelite House
50 Victoria Embankment
London EC4Y 0DZ

www.hoddereducation.com

Impression number   10 9 8 7 6 5 4 3 2 1

Year   2026 2025 2024 2023 2022

Cover photo © Satit_Srihin – stock.adobe.com

Illustrations by Integra Software Services Pvt. Ltd., Pondicherry, India

Typeset in Integra Software Services Pvt. Ltd., Pondicherry, India

Produced by DZS Grafik, Printed in Bosnia & Herzegovina

A catalogue record for this title is available from the British Library.

Contents

How to use this book

To make your study of Cambridge Checkpoint Science as rewarding as possible, look out for the following features when you are using this book:

● These aims show you what you will be covering in the chapter.

Do you remember?

This will show you the ideas you have learnt before. Think about what you already know before you begin.

Science activity

Science activities may be about developing a science skill or making a science enquiry.

Science in context

In this box you will find information about how scientists working alone or together have built up our understanding of the world over time, how science is applied in our lives, the issues it can raise and how its use can affect our global environment.

Science extra
The information in these boxes and any other boxes which have the 'Science extra' heading is extra to your course, but you may find these topics interesting and they may help you with your understanding of the overall chapter topic.

DID YOU KNOW?
This is a fact or piece of information that may make you think more deeply about the topic, or that you may share as a fun fact with your family and friends.

Summary

This box will show you how much you have learnt by the end of the chapter.

This book contains lots of activities to help you learn. Some of the questions will have symbols beside them to help you answer them.
Look out for these symbols:

 This blue dot shows you that you have already learnt some information to help you with this topic.

 If a question has a purple link symbol beside it, you will have to use your skills from another subject.

 This star shows where your thinking and working scientifically enquiry skills are being used.

 Scientists use models in science to help them understand new ideas. This icon shows you where you are using models to help you with your ideas in science.

 This icon tells you that content is available as audio. All audio is available to download for free from www.hoddereducation.com/cambridgeextras

 There is a link to digital content at the end of each unit if you are using the Boost eBook.

CHALLENGE YOURSELF

These activities are a challenge! You may have to think a bit harder to get the correct answer.

LET'S TALK

When you see this box, talk with a partner or in a small group to decide on your answer.

Work safely

This triangle provides you with extra guidance on working safely.

Words that look like **this** are glossary terms, and you will find definitions for them in the glossary at the back of this book. Other key terms that may not be included in the glossary look like **this**.

Introducing science

In this chapter you will learn:
- how to describe and explain in science
- how to continue to think and work scientifically
- how to look closely at scientific enquiry
- how to use models and analogies in science
- how to think creatively
- how to thought-shower
- how to move from measurement to calculation.

Do you remember?

- How do we think the solar system formed?
- Describe the structure of the Earth.
- What are the main parts of an animal cell?
- How can you tell an animal cell from a plant cell?
- Name five forms of energy.
- What do you use an ammeter for?

Describing and explaining in science

You should be able to answer all the questions in the *Do you remember?* section from the work you did last year. The answers to these questions are quite short, but can you answer longer questions which need more scientific description and explanation? Here are some to try.

▲ **Figure 1** In Figure 1a a Bunsen burner is shown, but you may have used a spirit burner as shown in Figure 1b.

1 How would you set up the equipment to heat a liquid, and how would you take its temperature?

DID YOU KNOW?

The first people to really begin to have ideas about the world were the Ancient Greeks. They spent a lot of time arguing about what they thought!

▲ Figure 3

2 How would you test a liquid to find out if it was an acid or an alkali? What results would you look for?

◀ Figure 2

3 How would you set up a microscope to look at a specimen on a slide?

Thinking and working scientifically

Science enquiry

The store of scientific knowledge has been built up by making scientific enquiries. A scientific enquiry is divided into three stages, and in each stage there are five or more scientific activities, as shown starting below.

▲ **Figure 4** Collaboration can be important when setting up a scientific enquiry.

Setting up an enquiry

- Look at a **hypothesis** and think about whether it can be tested.
- Think about how the evidence for an enquiry could support or contradict a hypothesis.
- Use your scientific knowledge and understanding to make a prediction of what might happen in an enquiry.
- Use a range of types of investigations in your planning of enquiries, remembering that they do not all involve fair tests, and think about the variables you need to consider as you plan.
- As you plan, think about the risks you may encounter and make a risk assessment to control them.

Scientific enquiry: purpose and planning

Carrying out an enquiry and recording data

- Use your classification skills in testing and observing organisms, materials, objects and phenomena, such as biological, chemical and physical processes.
- Select equipment for your enquiry and use each piece properly and safely.

▲ **Figure 5** Knowing how to use equipment and to record data accurately is an essential skill.

▲ **Figure 6** Scientists communicate their findings to peers, which helps to further scientific discovery.

- Decide to make a small number of observations or measurements (perhaps three) and record them. Look through them to see if they differ or are the same. If they differ make more observations and measurements until you think the data is reliable and can be used to test your hypothesis.
- Use all measuring instruments to provide accurate and precise measurements, explaining why this is necessary.
- Work safely at all times and follow the guidance in risk assessments that have been made.
- Use information from a range of secondary sources, such as books and the internet, but evaluate them with care to make sure they are relevant to your enquiry and are not providing biased information.
- Collect enough observations and measurements to make your data reliable, and record them in a form that will be easy to analyse and evaluate.

Examining the results and drawing a conclusion

- Compare the predictions with the results of the enquiries and assess their accuracy.
- Look through the results to see if they show a pattern or a trend, and note any. Look for results (known as anomalous results) which do not fit in with any pattern or trend and note them too.
- Look through the results, interpret what you see and draw a conclusion, but explain how the conclusion could be limited and might not provide a full answer to the enquiry.
- Review the investigations and experiments you have made in the enquiry and state how they may be improved. Explain why the changes you may have identified will provide more reliable data for the enquiry.
- Communicate your observations and measurements by presenting them clearly and providing a clear interpretation of what you think they show.

A closer look at scientific enquiry

There are many scientific terms featured in the discussion of the scientific activities above, and, to help you use these words when thinking scientifically and explaining your work to others, here are some descriptions of a few of them.

- **Questions –** scientific enquiries begin with questions. This is based on an observation and usually begins with words like *how, if* or *when*.

- **Hypothesis –** an idea to explain an observation. In science it must be a testable hypothesis. This means that the explanation can be tested by a scientific enquiry such as an experiment. Some hypotheses are untestable. Two examples of untestable hypotheses are 'red is better than yellow' and 'unicorns make the sky blue'.
- **Prediction –** a guess, using your knowledge and understanding, at what might happen when an experiment or investigation is carried out.
- **Investigation –** a process to find something out, which may or may not involve an experiment.
- **Experiment –** a process in an investigation when a condition or variable is changed to see how it affects other conditions or variables.
- **Variables –** features that are investigated in an experiment. There are three kinds: the independent variable (this is the variable or condition that the scientist changes in the experiment); the dependent variable (this is the variable or condition that changes as a result of changing the independent variable); the control variable (this is a variable that could affect the results of the experiment but is not to be investigated at this time, so it is kept constant or the same). In some experiments the effect of temperature would not be required so the temperature is kept constant and is controlled.
- **Fair test –** an experiment in which all variables are kept constant except the independent variable so that its effect on the dependent variable can be observed.
- **Trend –** the way data changes gradually from one value to another; for example, the numbers go up or down over time.

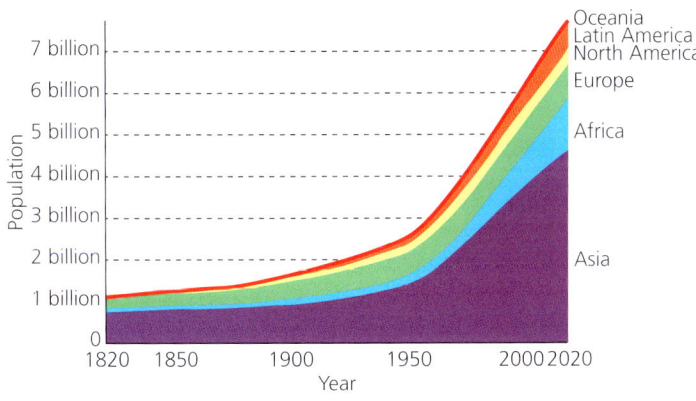

▲ **Figure 7** This graph shows an upward trend over time.

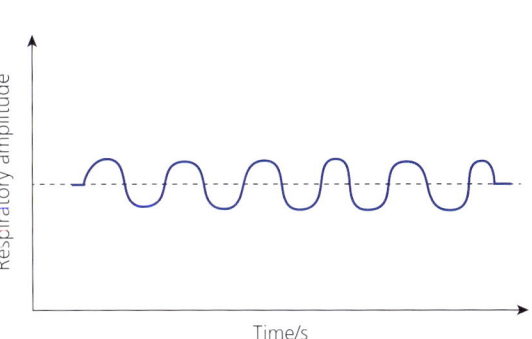

▲ **Figure 8** This graph shows a repeating pattern.

- **Pattern –** the way a set of data is repeated over time.
- **Anomalous result –** a result that does not fit into a trend or a pattern.
- **Data –** the collection of facts or measurements, collected during an investigation, which is analysed, evaluated and used to draw conclusions.

CHALLENGE YOURSELF

Using any symbols you can remember for drawing scientific diagrams, draw a diagram for the equipment being used in Figure 9.

▲ **Figure 9**

Thinking and working scientifically

Models and representations

Using models and analogies in science

When the data collected in a scientific enquiry is analysed and evaluated and conclusions are drawn, scientists sometimes make a model so that data and conclusions are easier to understand. In Grade 8 you should be able to:

- Know what an analogy is and how analogies can be used in making models.
- Use an analogy that is widely known in a scientific explanation.
- Represent some scientific ideas using symbols and formulae.

In science, a **model** is something that is used make the objects, systems and/or processes easier to understand. Using a model can help you make predictions or explain your observations.

Models can be physical or conceptual. Physical models are often 3-D representations, for example: a model of a circuit, or a model that uses balls to represent the rotation of Earth around the Sun. An aquarium in the laboratory could be used as a model of a pond.

Conceptual models are about *relationships* between the things being studied, for example, a model Earth orbiting the Sun can also be a conceptual model, because it shows the relationship between them. However, most conceptual models are drawings, diagrams, photographs, charts and graphs or equations.

It is important to think about and identify the limitations of any models used. You will be given opportunities to develop this skill throughout this student book.

▲ **Figure 10** Everyday items can be used for scientific analogies and models.

An **analogy** is the comparison of one thing (something unfamiliar) with something else (which is more familiar). The aim of the comparison is to make the unfamiliar thing easier to understand by thinking about something more familiar. The familiar thing is called the 'analog' of the unfamiliar thing. For example, a boiled egg can be used as a model for the structure of the Earth. The boiled egg is the model of the planet and the parts inside the egg are analogies for parts of the planet – the shell (crust), white (mantle), yolk (core). These analogies can be used to help you understand the structure of the Earth. This year we shall look at analogies more closely and assess their usefulness, and you will even make your own.

For example, a football and a table tennis ball can be used as analogies for the Sun and the Earth. They can be used to make a model of how the Earth goes around the Sun. In this model, someone holds up the football and a second person walks around them with the table tennis ball.

Creative thinking

4 Look at Figure 11. Who is thinking about a question and who thinks they may have made a discovery?

5 To begin a creative thinking exercise, it is important to get the brain to relax and think about connecting things which you would not normally link together. Does the temperature of a liquid affect the speed at which it flows?

6 Which property is being investigated in each of the tests 1–6?

You might think that **creative thinking** is all about writing poetry or painting a picture, and that it is only to do with arts subjects, not science. However, creative thinking is very important in science too, as it allows scientists to think freely about a question and come up with lots of suggestions which could be tested. Sometimes creative thinking helps them to find links between ideas and information, and this helps scientists to set up a **scientific theory**.

▲ **Figure 11** At any time, scientists can have a creative thought which could help them in their discoveries, just like these students, as they prepare to study in the laboratory and out in the environment.

To be a creative thinker you need to have a lot of things to think about, so the more you research and find out, the better your chances of thinking creatively. Imagine that you have been given two blocks made of different materials and have been asked to compare them. If you think about the properties of materials which you investigated last year, you may come up with the following series of tests.

1 Bend each block.
2 Scratch the corner of one block on the flat surface of the other to see which makes the deeper mark.
3 Hit each block with a hammer to see which one is shaped more by the impact.
4 Put the same amount of water on the top of each block and look to see whether any goes inside the block.
5 Put each block in warm water and, after a few minutes, test whether one feels warmer than the other.
6 Put each block into a simple electrical circuit with a battery, switch and lamp, and find out whether either of them conducts electricity.

All of these ideas may come to you from your work last year, but how would you find out which block was the shiniest? You may have to think harder.

You could begin by thinking about which source of light to use and you might conclude that a flashlight would be the best thing. You could then shine it on the same area of flat surface of each block to make the test fair and to see which block shines more.

In science investigations, however, something is usually measured, so how could you measure the shininess of each block?

Your creative thought might flow in the following way.

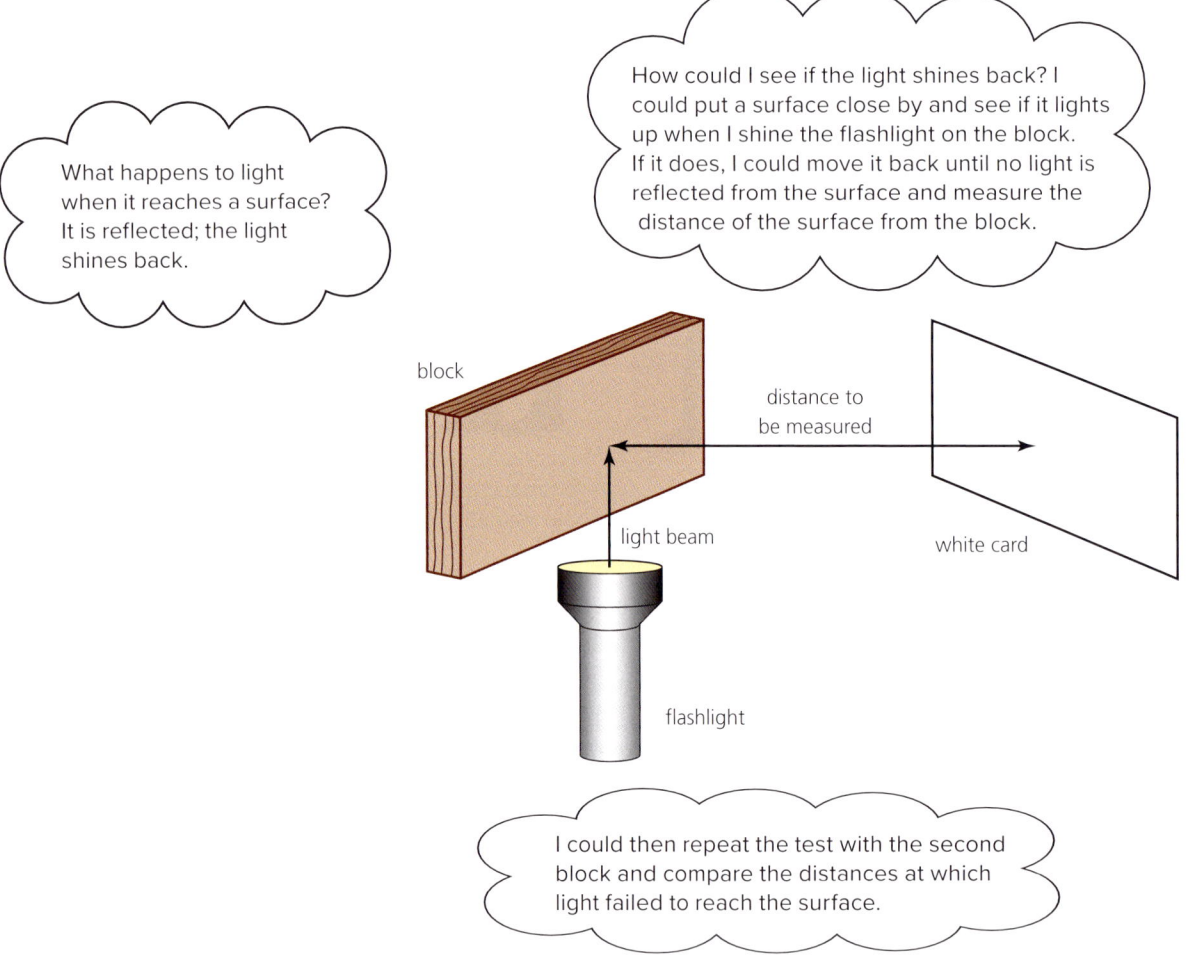

What happens to light when it reaches a surface? It is reflected; the light shines back.

How could I see if the light shines back? I could put a surface close by and see if it lights up when I shine the flashlight on the block. If it does, I could move it back until no light is reflected from the surface and measure the distance of the surface from the block.

block

distance to be measured

light beam

white card

flashlight

I could then repeat the test with the second block and compare the distances at which light failed to reach the surface.

▲ **Figure 12** Equipment for measuring reflection.

7 How can you use the measurements to show which block is shinier?

LET'S TALK

Do people memorise facts better in silence or when they listen to music? How could an investigation on this topic be carried out? Thought-shower in a group, then build an investigation to find out.

Make a prediction, then carry out your investigation and present your results and conclusions to the rest of the class.

8 Does the data in the table support or contradict this hypothesis: 'The length of every pace is always the same as you walk along'? Explain your answer.

9 What equipment was used in the investigation?

10 a What risks could there be to the safety of the walker if the investigation took place outside?

   b How could you make sure the investigation was safe to carry out?

11 How was the data made more reliable?

12 How was the data presented for examination?

Thought-showering

A group of between five and seven people can take part in a creative-thinking exercise called **thought-showering**. A problem is given to the group and everyone makes suggestions about how it could be investigated. It is just an idea-collecting exercise, and one person should write down the ideas as they occur. Any idea can be submitted, even the first idea someone thinks about. It does not matter if it seems silly, as it might give someone else another more serious idea. Once a list of suggestions has been collected, the thought-showering can be stopped and the group can then select ideas from their list to test.

From measurement to calculation

The measurements collected in an enquiry, in order to provide information that answers one question, can sometimes also be used to answer other questions by performing a calculation on them. Here is an example.

If you walk 20 paces, do you always cover the same distance?

The plan for this investigation is to walk 20 paces and measure the distance covered with a long tape or a metre rule. This activity is repeated twice more.

The measurements are recorded in a table.

▼ Table 1

Walk	1	2	3
Distance covered/m	15.5	15.3	15.7

When the data in the table is examined, it reveals the answer to the question – that the same distance is not covered every time.

This discovery leads to another question: how much difference is there between the longest distance and the shortest distance? This in turn leads to a simple calculation: the subtraction of the longest distance from the shortest, which is $15.7 - 15.3 = 0.4\,m$.

The examination of the data also leads to a third question: how can the average (mean) length of pace be worked out from the data? The answer is revealed by doing a slightly more complex set of calculations.

The average (mean) distance is found by adding together the three distances covered:

$15.5 + 15.3 + 15.7 = 46.5\,m$

Then dividing the total by 3:

$46.5 \div 3 = 15.5\,m$

Then converting the average (mean) distance from metres to centimetres:

$15.5 \times 100 = 1550\,cm$

The average (mean) length of pace is found by dividing the average (mean) distance in centimetres by 20:

$1550 \div 20 = 77.5\,cm$

Summary

- ✔ Facts and activities in science can be communicated by describing and explaining.
- ✔ The Science enquiry activities in thinking and working scientifically can be divided into three sections.
- ✔ There are a number of scientific terms which are used regularly when talking or writing about science.
- ✔ Models and analogies can help us to understand scientific discoveries.
- ✔ Creative thinking is used in devising investigations.
- ✔ Investigations can be planned by thought-showering.
- ✔ Calculations of measurements can be used to answer questions.

End of chapter questions

 1 Does the length of your pace change when you change from walking to jogging? Plan an investigation to find out, and be sure to make a risk assessment, state your equipment and collect data you think is reliable. If your teacher approves, try it.

 2 How could you modify your plan in the answer to Question 1 to find out if a person's pace changes when they change from jogging to walking? If your teacher approves, try it.

1 Joints and muscles

In this chapter you will learn:
- to identify hinge joints and ball-and-socket joints
- about X ray photographs of joints (Science in context)
- how antagonistic muscles move the bones at a hinge joint.

LET'S TALK

What are the most common bones that people break? Do a survey in class to see who has broken a bone. Extend this to thinking about family and friends to try to provide a fuller answer. If someone has broken a bone and feels they can talk about it, ask how they broke it, how was it repaired and how long it took to heal.

Do you remember?

- State three important functions of the skeleton.
- Point to these bones in your body – skull, jaw, rib cage, hip, spine, a leg bone, an arm bone.

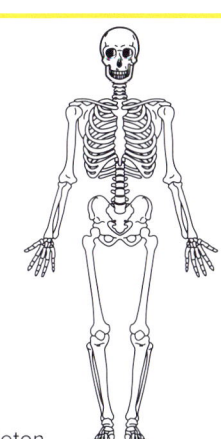

▶ **Figure 1.1** A human skeleton.

▲ **Figure 1.2** Trying out some equipment in a gym.

- How can you tell a muscle from a bone?
- How do pairs of muscles make bones move?

In this chapter we are going to look at types of joints which allow bones to move and how muscles move bones in one type of joint – the hinge joint.

Joints

The place where bones meet is called a **joint**. In some joints, such as those in the skull, the bones are fused together and cannot move. Most joints, however, allow some movement. Some joints, such as the elbow or knee, are called **hinge joints** because the movement is like the hinge on a door. The bones can only move forwards or backwards. A few joints, such as the hip and shoulder joints, are called **ball-and-socket joints** because the end of one bone forms a round structure, like a ball, that fits into a cup-shaped socket. This allows movement backwards and forwards, from side to side and even circular movement, as when you move your arm in a circle.

Two examples of the hinge joint are the elbow joint and the knee joint. Two examples of the ball-and-socket joint are the hip joint and shoulder joint.

DID YOU KNOW?

The strongest bone in the body is the femur. It is between your hip and knee joint.

1 Use Figure 1.3 to find each marked joint on your own body. Can you tell which type of joint each one is by feeling how they move? Explain your answer.

2 Label each letter shown on Figure 1.3 as either a hinge joint, or a ball-and-socket joint.

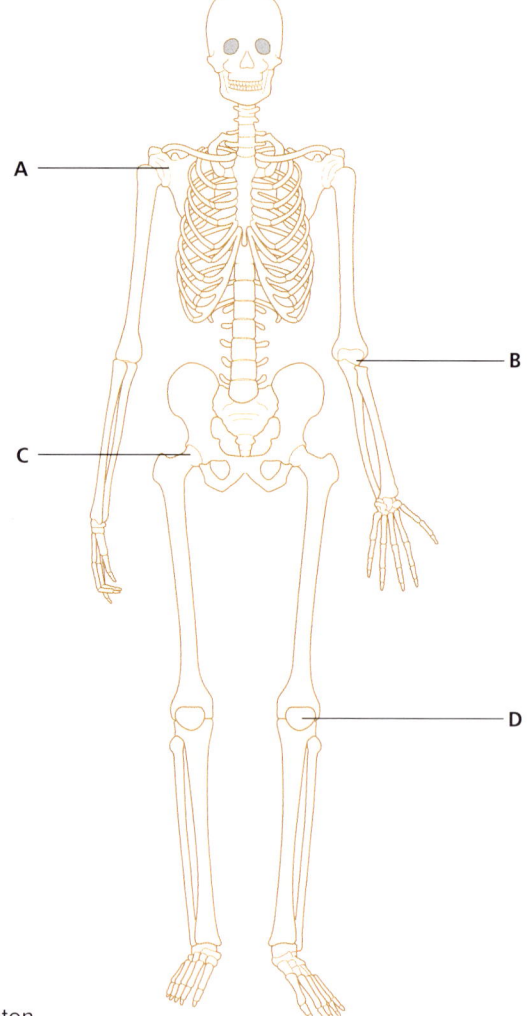

▶ **Figure 1.3** The human skeleton.

CHALLENGE YOURSELF

How are 3-D printers used to help people with damaged hands?

Survey the internet. Look at a few different sources. Do they all present their information in the same way, or are some biased or in favour of a particular point of view? Write a short account of three paragraphs or make a presentation of your findings.

This activity links well to thinking and working scientifically when you evaluate secondary sources of evidence for possible bias. It also links well to Science in context – how modern technology in 3-D printers can be used to address a real-life problem.

Science in context

X ray photographs of joints

When a joint is damaged through illness such as arthritis, or injury such as can occur in sport, doctors investigate by taking X ray photographs of the joint to help them plan a course of treatment.

X rays are a form of energy that can pass through flesh such as muscles and skin, but they are stopped by the material in bones, and this is what makes bones visible on X ray photographs.

3 Why would a doctor use X rays to investigate first, rather than cut open the flesh around the joint to have a look?

4 Here are X ray photographs of four joints.

Which joints are
a hinge joints and
b ball-and-socket joints?

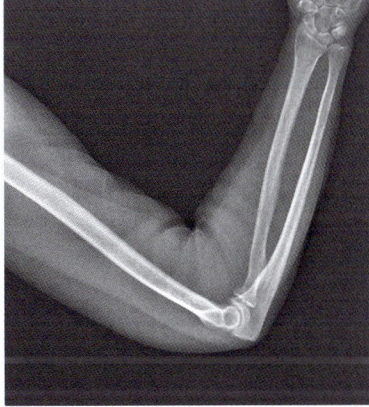

▲ **Figure 1.4 A** Photograph showing an X ray of an elbow joint

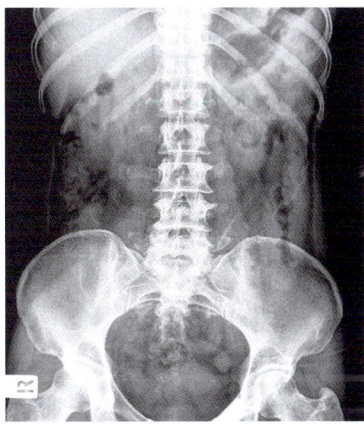

▲ **Figure 1.4 B** Photograph showing an X ray of a hip joint

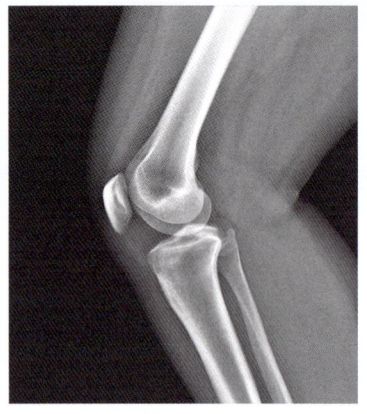

▲ **Figure 1.4 C** Photograph showing an X ray of a knee joint

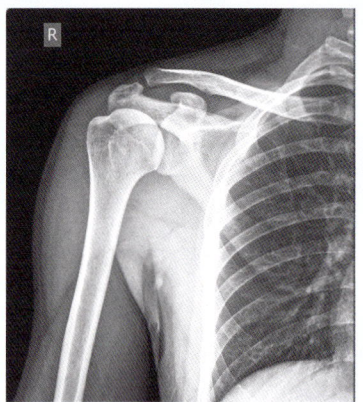

▲ **Figure 1.4 D** Photograph showing an X ray of a shoulder joint

Muscles in a hinge joint

Exercise the muscles in your upper arm to move your lower arm as shown in Figure 1.5.

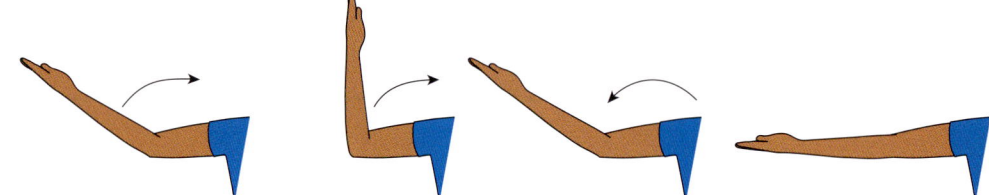

▲ **Figure 1.5** The muscles in your arm allow the hinge joint of your elbow to move like this.

The muscle action is producing movement across a hinge joint. To understand this action we need to look more closely at what is going on under the skin. Muscle is made up from tissue that has the power to move. It can contract to become shorter. A muscle is attached to two bones across a joint. When muscle gets shorter, it exerts a pulling **force**. This moves one of the bones, but the other stays stationary. For example, the biceps muscle in the upper arm is attached to the shoulder blade and to the radius bone in the forearm. When the biceps shortens (or contracts), it exerts a pulling force on the radius and raises the forearm.

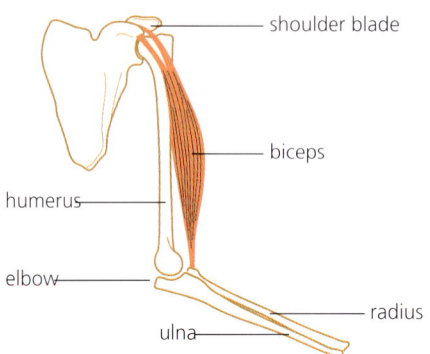

▲ **Figure 1.6** Biceps on arm bones.

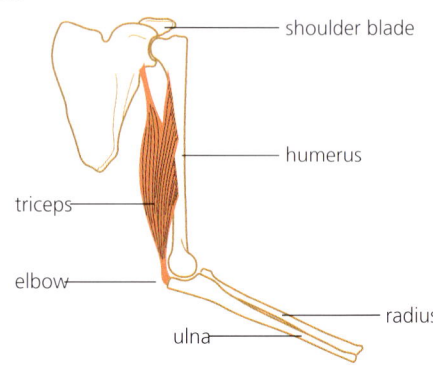

▲ **Figure 1.7** Triceps on arm bones.

5 Draw a diagram featuring both the biceps and the triceps, showing the triceps fully shortened.

6 Using dotted lines, draw on the position of the forearm when the biceps is fully shortened.

A muscle cannot lengthen or extend itself. It needs a pulling force to stretch it again. This force is provided by another muscle. The two muscles are arranged so that when one contracts it pulls on the other muscle, which relaxes and lengthens. For example, in the upper arm, the triceps muscle is attached to the shoulder blade, humerus and ulna. When it contracts, the biceps relaxes and the force exerted by the triceps lengthens the biceps and pulls the forearm down. When the biceps contracts again, the triceps relaxes and the force exerted by the biceps lengthens the triceps again and raises the forearm. The action of one muscle produces an opposite effect on the other muscle and causes movement in the opposite direction. The two muscles are therefore called an **antagonistic muscle pair**. The action can be summarised as follows:

- Biceps contracts, triceps relaxes, lower arm raised.
- Biceps relaxes, triceps contracts, lower arm lowered.

DID YOU KNOW?

Muscles are attached to bones by tendons. They do not stretch or shorten when the muscles do.

DID YOU KNOW?

Many scientists believe that the strongest muscle in the human body is the masseter muscle (or jaw muscle) which you use when you bite your food.

CHALLENGE YOURSELF

How does a muscle feel when it contracts and relaxes?

Stand up and let your left arm hang down by your side. Spread out the fingers of your right hand and push them into your biceps muscle. Move your fingers around a little to feel the muscle. Raise your left forearm, keeping the upper arm still, and feel the muscle with your fingertips. Lower your forearm again and feel the muscle.

Describe any changes that you felt in the muscle.

DID YOU KNOW?

Muscles which move your bones are called **skeletal muscles**, but there are two more types of muscle – **smooth muscle** which moves food along your digestive system, and **cardiac muscle** which forms the heart and pumps your blood around your body.

Can you feel changes in the triceps muscle?

This activity is part of thinking and working scientifically. It takes you through the process of setting up and carrying out a scientific enquiry.

Scientists often use an observation from an activity to set up an investigation. Here is an example. The changes felt in the *Challenge yourself* about the biceps can be used to set up a hypothesis to investigate the triceps muscle.

Hypothesis

When you feel your triceps muscle and raise and lower your arm you should feel a change in the muscles.

Prediction

Use the results of the challenge to predict what might happen to the triceps muscle when you raise and lower your arm.

Planning your enquiry

What will you do to test the hypothesis and prediction?

Investigation and recording data

Write down how the triceps felt when you
a   raised the forearm and
b   lowered the forearm.

Examining the results

Compare your observations when the forearm is raised and lowered.

Conclusion

Compare the examination of the results with the hypothesis and your prediction, and draw a conclusion.

Is your conclusion limited in some way? Explain your answer.

What improvements could be made? Explain the changes that you suggest.

7   Think about what you found out about the biceps in the challenge and the triceps in the investigation. Did you find any patterns or trends? Explain your answer.

Modelling muscles

This activity provides you with a good opportunity to apply your thinking and working scientifically skills by creating and using a model as a representation of how the joints and muscles work together to make your arm move.

Figure 1.8 shows an idea for making a model of the arm using elastic bands for muscles.

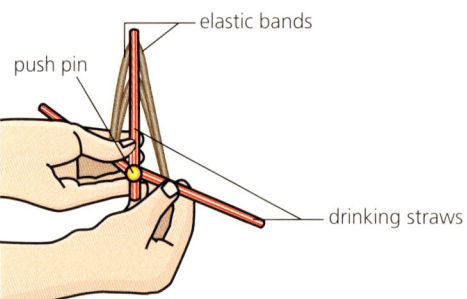

elastic bands
push pin
drinking straws

▲ **Figure 1.8** Model of an arm, using elastic bands for muscles.

Use the idea shown to make a model and demonstrate how the muscles work.

Here are some suggestions to help you make your model. You should discuss them with your teacher and perhaps search the internet for further ideas to make the model.

The bones could be made from straws or wooden craft sticks, and could be joined at the hinge with a carefully inserted drawing pin (thumb tack) which has its protruding point covered by a piece of sticky tack.

Thin elastic bands could be selected and held in place on the vertical stick by a notch cut into the stick by your teacher.

You may simply wish to try and hold the lower parts of the elastic band to the horizontal stick.

Describe the analogies used in your model.

What are the strengths and limitations of your model?

LET'S TALK

Look at the two sections on sports injuries in this chapter, then find out how frequent sports injuries are in your class. Make a list of injury types and how they happened, and then discuss what was done to help people recover from them.

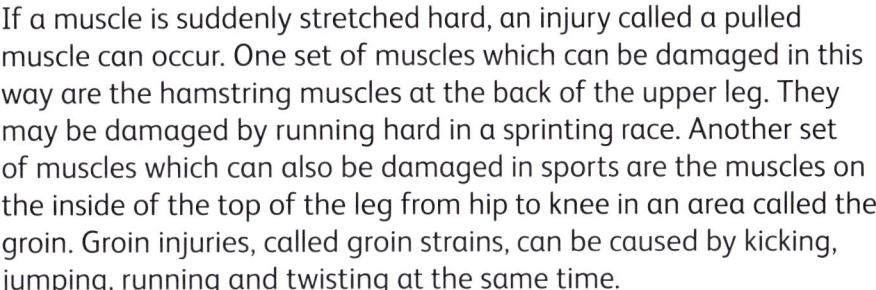

Science extra: Muscles and sports injuries

If a muscle is suddenly stretched hard, an injury called a pulled muscle can occur. One set of muscles which can be damaged in this way are the hamstring muscles at the back of the upper leg. They may be damaged by running hard in a sprinting race. Another set of muscles which can also be damaged in sports are the muscles on the inside of the top of the leg from hip to knee in an area called the groin. Groin injuries, called groin strains, can be caused by kicking, jumping, running and twisting at the same time.

CHALLENGE YOURSELF

Use the internet to find out about the common sport injuries in tennis, athletics, cricket, football and any other sport of your choice. Are some injuries found in more than one sport? Explain your answer. This is a good opportunity for you to examine the sources of evidence you choose. Can you identify any bias in any of them?

Summary

✔ Hinge joints, such as elbows and knees, move like the hinge on a door.
✔ Ball-and-socket joints, such as shoulders and hips, allow more circular motion.
✔ Science in context: X rays can be used to take photographs of joints.
✔ Antagonistic muscles move the bones at a hinge joint.

End of chapter questions

1 Name two hinge joints.
2 Name two ball-and-socket joints.
3 How is a hinge joint different from a ball-and-socket joint?
4 When the biceps contracts, what happens to
    a the triceps?
    b the lower arm?
5 What do you understand by the term 'a pair of antagonistic muscles'?

Now you have completed Chapter 1, you may like to try the Chapter 1 online knowledge test if you are using the Boost eBook.

2 Blood

In this chapter you will learn:

- about the parts of your blood and what they do
- about red blood cells and that they transport oxygen
- about sickle cell anemia (Science in context)
- about white blood cells and that they protect against harmful pathogens
- about plasma and that it transports blood cells, nutrients and carbon dioxide
- about glucose (Science extra)
- about saving lives with blood (Science in context).

Do you remember?

- What is the circulatory system?
- What travels around the circulatory system?
- Name some animals that have a circulatory system similar to yours.

When was the last time you saw your own blood?

Hopefully it was no more than a small cut or a graze, which everybody gets from time to time. From observations of our wounds we would say that blood is a red liquid, but it is much more. In this chapter we will look at what is in the blood and what the different parts of the blood do.

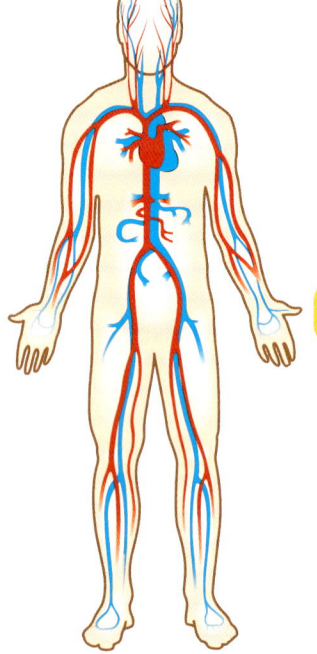

▲ **Figure 2.1** The heart and major arteries and veins.

▲ **Figure 2.2** A minor cut or graze.

▲ **Figure 2.3** You can see centrifugal force in action by swinging a bucket of water around quickly.

You may have tried the activity, shown in Figure 2.3, of putting some water in a bucket and whizzing it around you.

The water does not fall out of the bucket and soak you because of an effect called the centrifugal force. From this observation, scientists built a piece of equipment, called a **centrifuge**, which can spin liquids in test-tubes very fast.

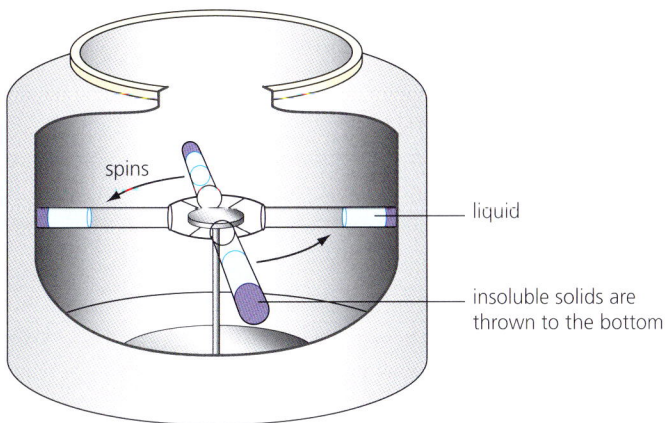

▲ **Figure 2.4** A centrifuge.

When a test-tube of blood is spun in a centrifuge, the centifugal force pushes the most solid materials to the bottom and the liquid settles at the top, as Figure 2.5 shows.

If you were to use a special stain on a drop of blood and then examine it under the microscope, you would see two types of cells – a large number of red blood cells, and a few heavily stained white blood cells, as shown in Figure 2.6.

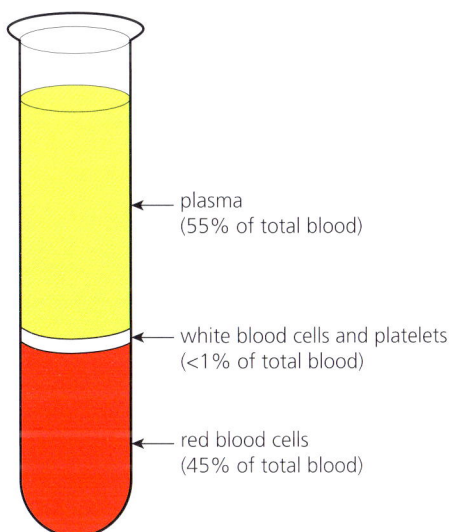

▲ **Figure 2.5** The components of blood.

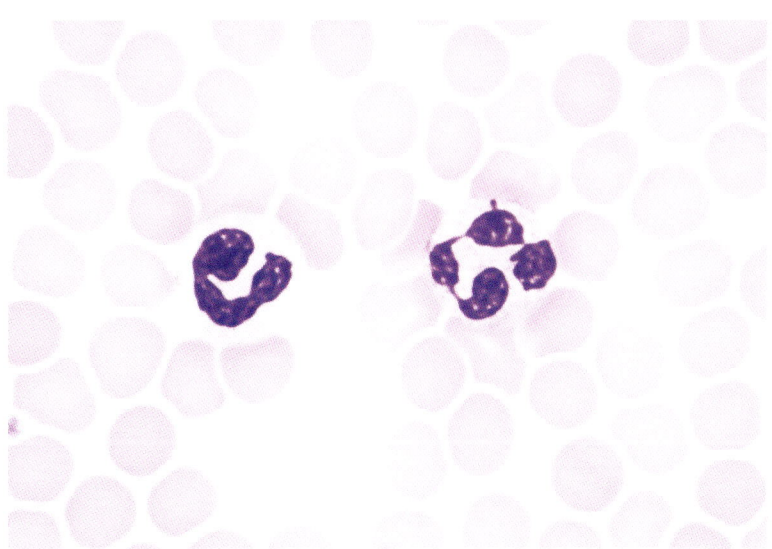

▲ **Figure 2.6** Blood viewed under a microscope.

▲ **Figure 2.7** A red blood cell.

1 Which part of the model represents the capillary?

2 What colour are the red blood cells when they come from the other cells in the body back to the lungs?

3 What colour are the red blood cells when they leave the lungs to go to the other cells in the body?

4 As the blood goes round and round the body, why does it change colour?

Red blood cells

Red blood cells collect at the bottom of the test-tube that is spun in the centrifuge. You can see in Figure 2.6 on the previous page that the red blood cells have a definite shape, and some of them seem lighter in the middle. The reason for this difference is that the centre of the cell, seen from the top, is thinner compared to the rim, which is thicker. Scientists call this shape a **biconcave disc**. Figure 2.7 shows just one side of a red blood cell with its rim raised above its centre. The other side has the same appearance.

DID YOU KNOW?

The human body makes over a million red blood cells every second in the sternum, ribs and vertebrae of the skeleton. They travel round the body for up to 120 days before they are destroyed in the liver and in an **organ** called the spleen.

A **capillary** is a small tube which takes blood through the tissues of cells. In the smallest capillaries, the red blood cells line up one after another to move through it.

Modelling red blood cells in a capillary

You will need:

red modelling clay, a clear plastic sheet which can be cut up and rolled to make a tube, sticky tape to hold the tube together, and scissors.

Process

1 Make a tube from the plastic sheet and sticky tape.
2 Make model red blood cells (biconcave discs) to fit inside it.
3 Keep your model for the *Challenge yourself* on page 13.

Red blood cells transport oxygen from the lungs to all parts of the body. When they are transporting oxygen they are bright red and when they have released their oxygen to other cells in the body they are dark red.

DID YOU KNOW?

The chemical in the red blood cell that makes it appear dark red is called **haemoglobin**. When the cell passes through the lungs it combines with oxygen to make **oxyhaemoglobin**, which makes the cell bright red.

Science in context

Sickle cell anemia

There is a blood disorder called sickle cell anemia, which is inherited and is not contagious (passed between people). When the red blood cells release their oxygen to other cells in the body, they change their shape to a sickle shape (like a crescent), stick together and can block small blood vessels. This causes pain, and over time can lead to damage of the liver, heart, lungs and brain, and an early death.

The cause of the disorder was identified in the early twentieth century by using microscopes to check blood samples of people with the condition. The disorder is controlled by the use of painkillers, as the only cure is to have a bone marrow transplant and this is not possible in many cases.

Most of the people who have this disorder live in Africa, below the Sahara desert. In Nigeria, around 2% of newborn babies have the disorder, and at the University of Lagos, research led by Dr Joy Okpuzor is taking place to help them. This research is focused on the substances produced by the leaves of the moringa, or drumstick tree, and it is hoped that the results will help people with the condition.

CHALLENGE YOURSELF

How widespread is sickle cell anemia around the world? Use online maps to find out. Is your country affected? If it is, research further to find out what is being done to help people with the disease.

White blood cells

These form in the thin, light-coloured layer in the test-tube that was spun in the centrifuge (see Figure 2.5). White blood cells do not have a fixed shape like red blood cells. Their shape is described as an **irregular** shape. In fact, they keep changing their shape as the cytoplasm flows about inside them.

White blood cells defend the body against disease-causing microorganisms called **pathogens**. Harmful bacteria and viruses are two examples of pathogens. They may enter the body through a cut in the skin, or through the moist surfaces in the nose and mouth, and the inner surfaces of the **lungs** and **digestive system**.

White cells destroy pathogens. They gather in large numbers wherever the pathogens have entered the body. They destroy them by eating them, as Figure 2.8 shows.

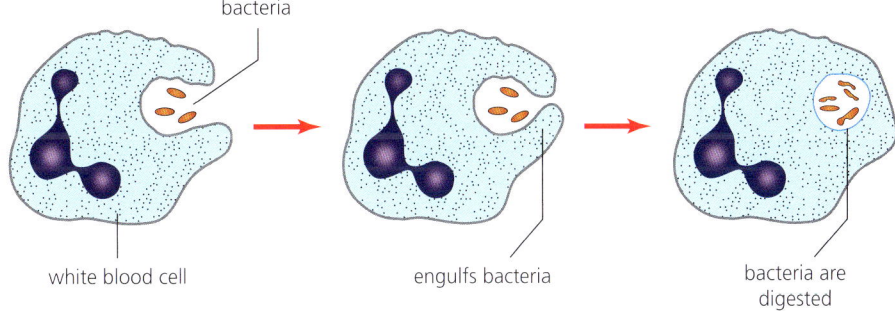

bacteria

white blood cell   engulfs bacteria   bacteria are digested

▲ **Figure 2.8** A white blood cell eating bacteria.

When a white blood cell has finished eating pathogens, it dies. If the pathogens have entered by a cut in the skin, a large number of dead white cells collect around the wound and form a substance called **pus**. As the amount of pus builds up to kill off the pathogens, a scab forms to keep more pathogens from entering the wound.

LET'S TALK

As you were growing up, there is a good chance that you had wounds caused by accidents that made scabs as your white blood cells attacked the invading pathogens. Think back and see if you can remember them. Did everyone develop scabs in the same places when they were children?

Modelling a white blood cell attack

Process

1 Use Figure 2.8 as a guide to make your model.
2 What materials will you use? How big will you make your model?
3 Draw up a plan and, if your teacher approves, make it.
4 Keep your model for the *Challenge yourself* on page 13.

5 Our blood has a certain number of white cells in it, ready for attacking pathogens. Sometimes this number goes down due to ill-health. Does a reduced number of white cells in the blood make it easier for pathogens to invade the body? Explain your answer.

DID YOU KNOW?

Some white blood cells make substances called **antibodies**, which stick to pathogens to make them easier for other white blood cells to attack.

Plasma

This is the straw-coloured watery liquid shown in Figure 2.5 (page 9). It transports the red blood cells and the white blood cells around the body. It also contains a range of substances dissolved in it which it transports too, such as carbon dioxide in small amounts. Plasma also contains **nutrients** absorbed from the digestive system, such as glucose, vitamins, fats and minerals, which are important in cell function.

Science extra: Glucose
Glucose is one of the nutrients found in plasma. It forms when carbohydrates (see page 26) are digested and it takes part in respiration in the cells of the body, which you will study in the next chapter.

Make a presentation about blood, using your models from earlier in the chapter and any other information you may have. You could begin with a 'Did you know?' based on the following information:

- A newborn baby has about 270 cm³ of blood in its body, a child has about 2650 cm³ and an adult has between 4500–5700 cm³.

You could use containers of water, coloured with red food dye, to illustrate these facts and communicate them to the class, then continue with your own presentation, or even make a video.

Science in context

Saving lives with blood

People with injuries, diseases, or those recovering from operations sometimes need to be given blood to replace any they have lost. This blood is provided by an adult person called a **blood donor**. A blood donor may give about 470 cm³ of their blood at one time. They may make up to four donations a year. Over 100 million blood donations are made every year in countries across the world. There is even a World Blood Donor Day!

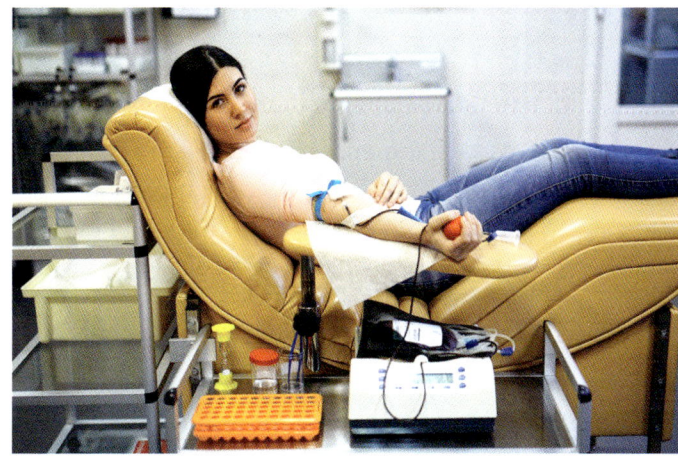

▲ **Figure 2.9** A person donating blood.

Each person's blood has a certain set of features. Some people have blood with similar features, but some people have different features in their blood. If a person is to receive blood, it has to be similar to their own, so great care is taken to match the donor's blood to the person receiving it.

The donated blood is treated to preserve it and is stored in a blood bank, ready for use. When it is needed, it is transported to hospitals or emergency sites and is released into the patient in a process called a **blood transfusion**.

6 When a person needs blood, why must it be delivered to them quickly?

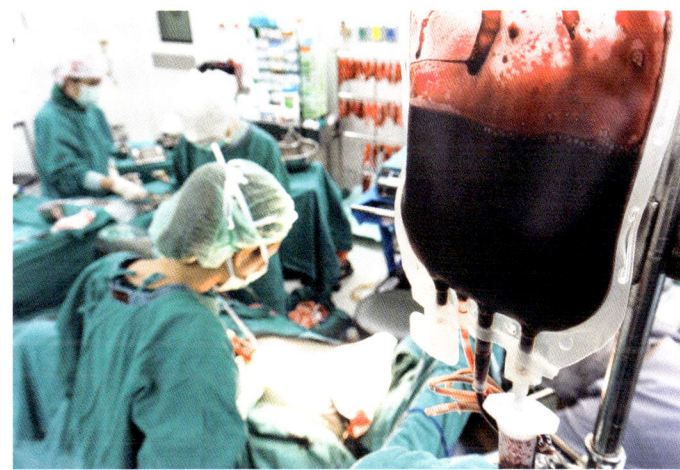

▲ **Figure 2.10** A person receiving a blood transfusion.

Summary

✔ Blood is made up of plasma, red blood cells, and white blood cells.
✔ Red blood cells transport oxygen.
✔ Science in context: Red blood cells are concave; in someone who suffers from sickle cell anemia they are shaped like a crescent.
✔ White blood cells protect against harmful pathogens.
✔ Plasma transports blood cells, nutrients and carbon dioxide.
✔ Science in context: Blood transfusions can be given to patients who need replacement blood.

End of chapter questions

1 Describe the appearance of a red blood cell (not just its colour).
2 What does a red blood cell carry round the body that cells need?
3 Describe the appearance of a white blood cell (not just its colour).
4 What task do white blood cells perform to keep the body alive?
5 What is the plasma and what does it do?
6 Imagine you are in a tiny submarine travelling in the blood.
   a When you look out of the window, what do you see?
   b What do you see when you come to a wound with a broken blood vessel?

 Now you have completed Chapter 2, you may like to try the Chapter 2 online knowledge test if you are using the Boost eBook.

In this chapter you will learn:
- the difference between breathing and respiration
- about summary word equations and how to use them to explain aerobic respiration
- about aerobic respiration in plant and animal cells and how it gives a controlled release of energy
- about anaerobic respiration (Science extra)
- how the respiratory system is structured to enable gas exchange
- about the windpipe (trachea) (Science extra)
- about asthma (Science in context)
- about the diffusion of oxygen and carbon dioxide between blood and the air in the lungs.

Do you remember?

- What gas do we need from the air to keep us alive?
- What do we do to take this gas into our bodies?
- What is the human respiratory system?
- What other vertebrates have a respiratory system that is similar to ours?
- How do you breathe when you have finished running a race as shown in Figure 3.1? (You can find an explanation on page 16).

▲ Figure 3.1

The terms **respiration** and **breathing** are often confused, but they do have different meanings.

Breathing

Breathing describes just the movement of air in and out of the lungs. Breathing can also be described as ventilation.

Respiration

Respiration is a series of chemical reactions, taking place inside cells, to release energy. This series of chemical reactions can be summarised and represented by the word equation that is shown on the next page.

1 What would happen if the energy was released in an uncontrolled way?

2 What are the life processes that take place to show that an organism is alive?

In a word equation, the reactants (the starting materials) are shown on the left of the arrow and the products (the materials that are produced) are shown on the right of the arrow:

glucose + oxygen → carbon dioxide + water

You can see that in this summary of respiration, oxygen is a reactant. This type of respiration is called **aerobic respiration**. Aerobic respiration takes place in the mitochondria of the cells in all plants and animals, including in all our own body cells. The energy is released in a controlled way so that it can be used by other parts of the cell to keep the cell and the body alive.

Science extra: Anaerobic respiration

This process occurs when the body cannot get enough oxygen for aerobic respiration to take place. For example, when you sprint, you cannot breathe fast enough to get the oxygen you need to release energy for your muscles. The body responds by releasing the energy through a different process called anaerobic respiration, and other substances are made. These need to be broken down by aerobic respiration, so when your sprint is over, you breathe in large quantities of air quickly to provide the oxygen that is needed.

Parts of the respiratory system

Modelling the respiratory system

You will need:

spongy material for the lungs, but you may use any other materials and objects to make the other parts of the system. Write down the objects or materials you use.

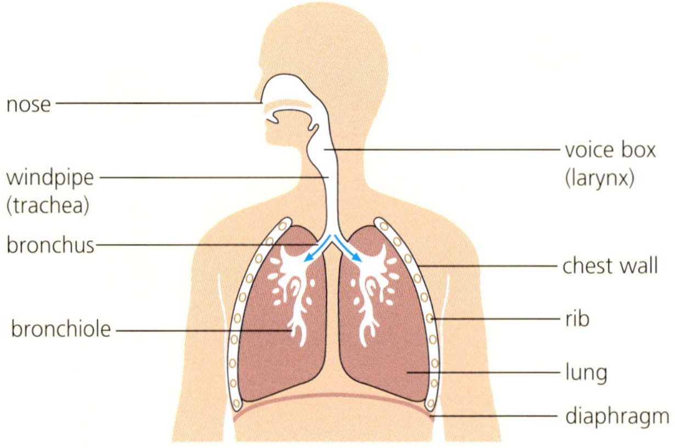

nose

windpipe (trachea)

bronchus

bronchiole

voice box (larynx)

chest wall

rib

lung

diaphragm

▲ **Figure 3.2** The respiratory system.

Process

Look at Figure 3.2. You will need to make a windpipe (trachea), a bronchus and interconnecting bronchioles in each lung, lung tissue, ribs and a **diaphragm**.

Refer to your model as you read through the rest of the chapter, then make a presentation about how the respiratory system works using your model.

If you take Figure 3.2 as the position once you have breathed in, you may like to make another model using Figure 3.7 (page 19) to show how the structures change when you breathe out.

The air passages and tubes

The nose

Air normally enters the air passages through the nose. Hairs in the nose trap some of the dust particles that are carried in the air currents. The lining of the nose produces a watery liquid called **mucus**. This makes the air moist as it passes inwards and also traps bacteria that are carried on the air currents. **Blood vessels** beneath the nasal lining release heat that warms the air before it passes into the lungs.

The windpipe (trachea)

The windpipe or trachea is about 10 cm long and 1.5 cm wide. It is made from rings of cartilage, which is a fairly rigid substance. Each ring in the windpipe (trachea) is the shape of a 'C'.

Science extra: The windpipe (trachea)

The inner lining of the windpipe (trachea) has cells which have microscopic hairs. These beat backwards and forwards to move the mucus (containing trapped dust and bacteria) to the top of the windpipe (trachea), where it enters the back of the mouth and is swallowed.

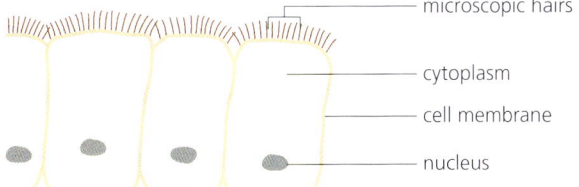

microscopic hairs

cytoplasm

cell membrane

nucleus

▲ **Figure 3.3** Microscopic hairs on cells in the windpipe (trachea).

The bronchi and bronchioles

The windpipe (trachea) divides into two smaller tubes called **bronchi** (singular: **bronchus**). The two bronchi are also made of hoops of cartilage and have the same lining as the windpipe (trachea).

The bronchi divide up into many smaller tubes called **bronchioles**. These have a diameter of about 1 mm.

3 What structures hold the air passages open in the windpipe (trachea) and bronchi?

The **chest wall** and the **diaphragm** surround the cavity in the chest. Most of the space inside the chest is taken up by the lungs.

The chest wall

This is made by the ribs and their muscles. Each rib is attached to the backbone by a joint that allows only a small amount of movement. The muscles between the ribs are called the **internal and external intercostal muscles**. The action of these muscles moves the ribs.

The diaphragm

This is a large sheet of muscle attached to the edges of the tenth pair of ribs and the backbone. It separates the chest cavity, which contains the lungs and heart, from the lower body cavity, which contains the stomach, intestines, liver, kidneys and female reproductive organs.

Science in context

Asthma

The bronchioles, shown in Figure 3.2 on page 16, have walls made of muscle and do not have hoops of cartilage like other tubes in the respiratory system. These wall muscles can make the diameter of the bronchioles narrower or wider. If a person has a disease known as **asthma**, their body responds to the presence of certain substances around them by making the bronchioles narrower. This makes breathing very difficult and the person is said to have an **asthma attack**.

Substances which can cause an asthma attack in some people are dust mites (that live in bedding and carpets) animal fur and bird feathers, pollen, mould spores, and chemicals that cause air pollution. A person who responds to these substances is said to be **allergic** to them.

Asthma is a disease that occurs all over the world in both children and adults. There are a range of treatments to reduce or prevent an asthma attack. A common treatment is the use of an **inhaler**, which contains a medicine that makes the muscles in the bronchioles relax and widen so that breathing becomes easier.

▲ **Figure 3.4** A person using an inhaler.

LET'S TALK

Ask the people in your group if anyone in it has asthma. Perhaps they would like to say what causes it and how they treat an attack, such as using an inhaler and following up with a peak flow meter. Some people may know of others who have asthma and know about what they do. How can sharing information about a disease stimulate someone to want to help?

4 Why is it more difficult to breathe during an asthma attack?

People who have asthma may also use a peak flow meter regularly to find out how far open their bronchioles are. They blow hard into the meter; if a high score is recorded, the bronchioles are wide, but if the score is low, the bronchioles are narrow.

Research into lung diseases like asthma is taking place all over the world. Sometimes a person goes into this research because of a member of their own family.

Dr Sundeep Salvi, the director of the Chest Research Foundation (CRF) in Pune, India, became a doctor and researcher after seeing his mother live with asthma.

▲ **Figure 3.5** A person using a peak flow meter.

▲ **Figure 3.6** Dr Sundeep Salvi, director of the Chest Research Foundation.

Breathing movements

There are two breathing movements: **inspiration** and **expiration**. The actions that take place during each are shown in Figure 3.7 and are summarised in Table 3.1.

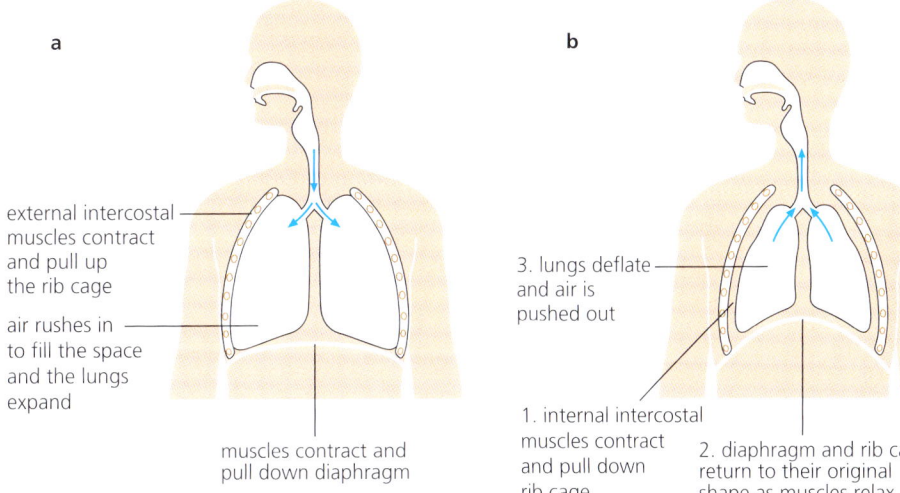

a

external intercostal muscles contract and pull up the rib cage

air rushes in to fill the space and the lungs expand

muscles contract and pull down diaphragm

b

3. lungs deflate and air is pushed out

1. internal intercostal muscles contract and pull down rib cage

2. diaphragm and rib cage return to their original shape as muscles relax

▲ **Figure 3.7a** Inspiration; **b** Expiration.

▲ **Figure 3.8** Feeling the movements made during breathing.

▼ **Table 3.1** Inspiration and expiration.

Part of respiratory system	Inspiration	Expiration
internal intercostal muscles	relax	contract
external intercostal muscles	contract	relax
ribs	move up	move down
diaphragm muscles	contract	relax
diaphragm	moves down and becomes flatter	moves up and becomes dome-shaped
chest volume	increases	decreases
air	moves in	moves out

5 How does the action of the external intercostal muscles and the diaphragm muscles draw air up your nose?

How do different experimental techniques compare?

Scientists try to develop different techniques to investigate an enquiry. In this investigation, you are to compare two techniques for finding out how much air you can blow out in one go.

When you have read the procedure for each technique, make a plan for how you will use it, for example, how many times will you repeat the technique and how you will record your data? After completing the two investigations, compare the data you have collected and draw a conclusion, stating how the conclusion could be limited.

Work safely

This is not a competition to see who can blow out the most air, so do not make it so. It is simply to compare techniques for measuring how much air you can blow out in one go.

● **Use a new or sterile balloon and mouthpieces and do not share them.**

● **Do not take part in this if you are asthmatic or have any other breathing issues.**

6 How could you improve this part of the investigation for technique 1 to make the data more reliable?

Technique 1

You will need:

a balloon, a bucket, a deep tray, a measuring jug.

Process

1 Place the bucket in the deep tray.
2 Fill the bucket to the brim with water.
3 Blow up the balloon in one breath.
4 Push the inflated balloon into the bucket of water so that it pushes water over the side and into the tray.
5 Empty the tray into the measuring jug and record the volume of water in the jug (if the jug is small, you may have to fill it several times).
6 The volume of water pushed out of the bucket is the same as the volume of air in the balloon, which is the volume of air you blew out.

Technique 2

You will need:

a 3-litre clear plastic bottle, a marker, a 200 cm^3 jug, a large plastic bowl and a clean, clear, flexible plastic tube.

Process

1 Pour 200 cm^3 of water into the bottle and mark its level.
2 Repeat step 1 until the bottle is full and you have a scale up the side of the bottle.
3 Fill a third of the bowl with water.
4 Close the top of the full bottle of water with your hand or a lid.
5 Turn the bottle upside down, keeping the top closed and insert it into the bowl of water.
6 Open the top of the bottle underwater and insert one end of the plastic tube into it.
7 Blow down the open end of the tube so that your air forces water out of the bottle.
8 Measure the water level in the bottle from the top and work out the volume of air in the bottle. This the same as the volume of air you blew out.

7 How could you improve this part of the investigation for technique 2 to make the data more reliable?
8 What do you conclude from the comparison of the techniques?

Does a person's breathing rate change?

You will need:

a stop-clock or timer.

Hypothesis

As breathing supplies oxygen to the body to release energy for activity, it seems reasonable to suppose that increasing activity increases the demand for oxygen, and in turn increases the rate of breathing.

Prediction

Make a prediction based on the hypothesis.

Plan, investigation and recording data

Plan an investigation which compares the breathing rate after resting and after running. Make these decisions in your plan:

1   How long should someone rest and run for?
2   How long should the breathing rate be measured for?
3   Who should count the breathing rates? The person resting, or the person running, or an observer? Explain your decision.
4   How many times should the investigation be repeated? Explain your answer.

Examining the results

Compare the data you have collected about breathing after resting and running. Did your data show a pattern? If so, describe it.

Did some of the data produce one or more anomalous results? If so, identify them and suggest why they may have occurred.

Conclusion

What do you conclude from examining the data? Did increasing activity show an increase in the rate of breathing? Was the hypothesis testable?

How could the investigation be improved? Give reasons for your suggestions.

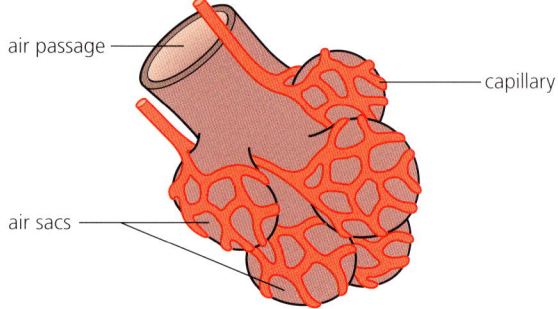

▲ **Figure 3.9** Air sacs.

Exchanging gases

At the end of each bronchiole is a short tube which leads to bubble-like structures called **air sacs**. Each air sac has a moist lining, a thin wall and is supplied with tiny blood vessels called **capillaries**. The air sacs form the **respiratory surface** and this is where a process called **gaseous exchange** takes place.

Oxygen from the inhaled air dissolves in the moist linings of the air sacs and moves by **diffusion** through the walls of the air sac and into the capillary next to it. Diffusion is a process in which a substance such as oxygen moves from a place where it is present in large quantities (the air in the lung) to a place where it is present in only small quantities (the blood in the capillary). You can find out more about diffusion on page 164.

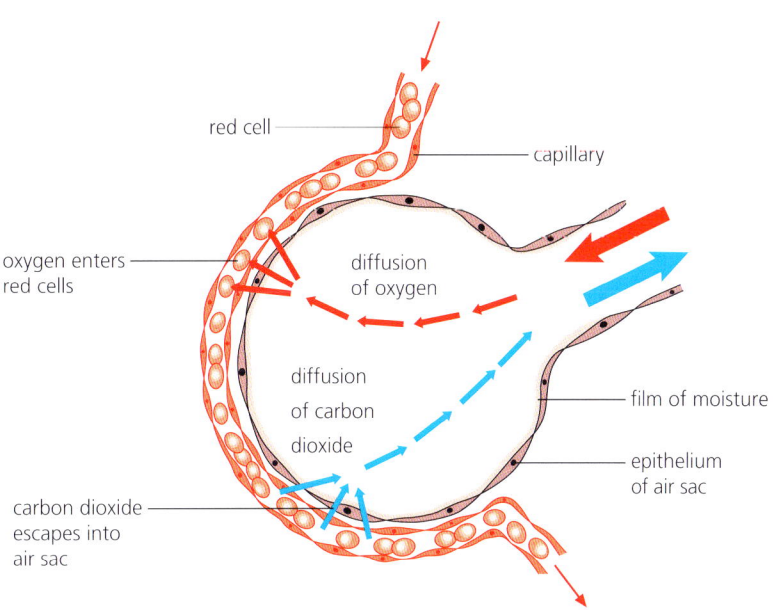

▲ **Figure 3.10** The direction of gaseous exchange.

9 Is the amount of carbon dioxide in the plasma greater than the amount in the air sacs? Explain your answer.

10 How would thick-walled alveoli affect the exchange of the respiratory gases?

The oxygen diffuses into the blood and enters the red blood cells, where it stays. Blood containing oxygen in the red blood cells is called **oxygenated blood**. The oxygenated blood is pumped away from the lungs by the heart.

We have seen that carbon dioxide is dissolved in the blood plasma. When it travels to the lungs, it moves by diffusion through the capillary walls, the air sac walls and linings and into the air in the air sacs, ready to be breathed out.

Blood moves through the capillaries very quickly, so a large amount of oxygen and carbon dioxide can be exchanged in a short time.

Explaining with an analogy

11 In the analogy
   a what does the surface area of the tennis court represent?
   b what do the footballs represent?

An **analogy** is when we say that something (A) is like something else (B) to help you to understand it. Sometimes the following analogy is used to help students understand the lungs and their surface area.

The lungs have a spongy structure, which is produced by their 300 million air sacs. They make a very large surface area through which oxygen and carbon dioxide can be exchanged (A). It is like the surface area of a tennis court being wrapped up inside two footballs (B).

LET'S TALK

How do you think you would be affected if the surface area of your lungs was reduced?

CHALLENGE YOURSELF

Red blood cells contain the mineral iron which the body gets from some foods. Iron helps the red blood cells take up and carry oxygen. How would eating a diet which lacked iron affect the amount of oxygen taken up by the blood?

Summary

- ✔ Breathing is the movement of air in and out of the lungs.
- ✔ Respiration is a series of chemical reactions inside cells leading to the release of energy.
- ✔ In a summary word equation, the reactants are shown on the left of the arrow and the products are shown on the right of the arrow.
- ✔ Aerobic respiration occurs in the mitochondria of the cells in all plants and animals.
- ✔ Science extra: Anaerobic respiration occurs when the body cannot get enough oxygen for aerobic respiration.
- ✔ The respiratory system consists of the nose, the windpipe, the bronchi and bronchioles, the chest wall, and the diaphragm.
- ✔ Science in context: Asthma causes the bronchioles of sufferers to narrow, making it hard for the person to breathe.
- ✔ The exchange of gases occurs through the diffusion of oxygen and carbon dioxide between blood and the air in the lungs.

End of chapter questions

1 What is the movement of air in and out of the lungs called?

2 What is the name of the process which is summarised in the following word equation?

glucose + oxygen → carbon dioxide + water

3 Where do you find mitochondria?

4 What happens in the mitochondria?

5 Describe the changes that take place in your chest when you

  a   breathe in

  b   breathe out.

6 Describe what happens in the lungs to the oxygen in the air and the carbon dioxide in the blood.

7 Do younger people have a faster resting breathing rate than older people? Plan an investigation to find out. Say what equipment you would use, and describe how you would make your measurements and control any risks.

Now you have completed Chapter 3, you may like to try the Chapter 3 online knowledge test if you are using the Boost eBook.

4   A healthy diet

In this chapter you will learn:
- that a balanced diet is made up of proteins, carbohydrates, fats and oils, water, minerals and vitamins in the right quantities
- about the functions of nutrients
- that carbohydrates and fats can be used as an energy store in animals
- that animals consume food to obtain energy and nutrients
- about the importance of fibre (Science extra)
- how to think and work scientifically to discover a vitamin (Science in context)
- about treating iodine deficiency (Science extra)
- about treating malnutrition and famine (Science in context).

Do you remember?

- Name the parts of the digestive system and describe what happens in them.
- What is a diet?
- Why do you need to think about your diet?
- What kind of diet is a healthy diet?

▲ **Figure 4.1**

▲ **Figure 4.2** A school canteen.

What kind of diet do you have? Here are two examples.

Diet 1
- No breakfast.
- Eat sweets on the way to school.
- Eat a chocolate bar and have a fizzy drink at morning break.
- Have fried food such as potato chips or fries with lunch.
- Eat some more sweets or candies in the afternoon.
- Avoid green vegetables in the evening meal.
- Have a snack of crisps (potato chips), sweets or candies and fizzy drinks during the evening.

Diet 2
- Have breakfast of cereal and milk, toast and fruit juice.
- Eat an apple at morning break.
- Have a range of foods for lunch during the week, including different vegetables, pasta and rice.
- Eat an orange in the afternoon.
- Eat an evening meal with green vegetables.
- Have a milky drink at bedtime.

1 Write a description of your daily eating pattern.

2 Compare your daily eating pattern with the two diets on the previous page. Which one does your daily eating pattern resemble?

3 From what you already know, try to explain which diet is healthier.

LET'S TALK

How healthy is your eating pattern? What changes would make it healthier? Do other people agree?

Nutrients needed by the body

A chemical that is needed by the body to keep it in good health is called a **nutrient**. The human body needs a large number of different nutrients to keep it healthy. They can be divided up into the following nutrient groups:

- carbohydrates
- fats and oils
- proteins
- vitamins
- minerals.

In addition to these nutrients, the body also needs water. Water accounts for 70 per cent of the body's weight and provides support for its cells. It carries dissolved materials around the body and helps in controlling body temperature.

Fibre is also needed by the body.

Carbohydrates

Carbohydrates can be used as a store of energy in animals, and animals must consume food to obtain them. There are many types of carbohydrate, but three of the most widely known are starch, sucrose (in table sugar, used in foods) and glucose (used in respiration). Starch is an energy store in plants and glucose is a chemical form of energy that travels easily in blood plasma to the cells of the body.

Plants also make another carbohydrate, called cellulose, which they use to make cell walls. It forms the part of our diet called fibre.

DID YOU KNOW?

The word 'carbohydrate' comes from the names of the two chemical **elements** that make it (carbon and hydrogen) plus the suffix '-ate', which means 'having oxygen' in chemistry.

DID YOU KNOW?

Chocolate contains a substance called theobromine which our bodies break down, making it harmless. However, dogs cannot break down theobromine, so eating chocolate could make them ill and possibly die.

◀ **Figure 4.3** The crunchy stalks of celery are made from cellulose.

DID YOU KNOW?
Fats are made from the elements carbon and hydrogen.

Fats

Fats can be used as a store of energy in animals, and animals must consume food to obtain them. There are two kinds of **fats**: the solid fats produced by animals, such as lard, and the liquid fats or oils produced by plants, such as sunflower oil. Fat is a store of energy in animals and is used as insulation beneath the skin to keep the bodies of mammals warm. Oils are energy stores in plants.

Many mammals increase their body fat in the autumn so that they can draw on the stored energy if little food can be found in the winter. Fats contain even larger amounts of energy than carbohydrates. The body cannot release the energy in fats as quickly as the energy in carbohydrates.

Food containing fat can leave a translucent (letting light pass through the material but scattering it in all directions) mark on paper and so can water. However, water evaporates from paper, but the fat in food does not. This information can be used to develop a fat test.

Science extra: Is it fat or water?

Work safely

Do not use the Sun as part of your translucency test, as it can permanently damage your eyes.

You will need:

a piece of absorbent paper, such as newspaper, a portion of butter in a container, a plastic knife, a small container of water, and a pipette or dropper.

Plan and investigation

Devise a plan with the following features:

1 a fair test to compare the size of marks of fat and water on the paper
2 a test to see if the marks are translucent
3 allowing a time and a place for the evaporation of water to occur
4 allowing an opportunity to check the presence of a fat mark.

Show your teacher your plan and, if approved, try it.

Science extra: Can food be tested for fat using paper?

Work safely

Hands should be washed thoroughly after this experiment.

You will need:

absorbent paper, such as newspaper, and a selection of foods including some that you normally eat.

Hypothesis

Marks made by rubbing food on paper can show when a food contains fat.

Plan, investigation and recording data

1 How many different foods will you test?
2 Look at each food in turn and predict whether or not it will contain fat.
3 Make a table in which to record your results.
4 You may think that some foods appear to have fat in them without trying the test. If so, make a prediction about them before you test them.
5 Carry out the fat test on each sample of food and record your observations in your table.

Examining the results

Analyse and evaluate the data in your table.

Conclusion

1 Draw a conclusion. In your conclusion
   a state if you thought the hypothesis was testable
   b assess your ability to predict which foods contain fat and give a reason for your answer.

Is your conclusion limited in some way? Explain your answer.

What improvements could be made? Explain the changes that you suggest.

Proteins

Proteins are found in both plants and animals. They are the materials that form most of the structures in our bodies, such as muscle and bone. Proteins are also needed for growth of the body and for repairing any damage due to illness or accident.

> **DID YOU KNOW?**
> Proteins are made from the elements carbon, hydrogen, oxygen and nitrogen. Some proteins also contain sulfur.

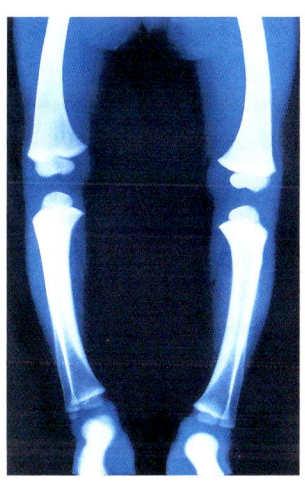

▲ **Figure 4.4** Hair is made of protein that grows out from your skin.

▲ **Figure 4.5** This child has rickets. It can be prevented by adding vitamin D to the diet.

4   In a patient with rickets, why do the leg bones bend more than the arm bones?

How the body uses nutrients

Science extra: Fibre

Cellulose is a carbohydrate which makes up the walls of plant cells. The cellulose in food is known as **dietary fibre**. It is found in foods such as wholemeal bread, fruit and vegetables. We cannot digest fibre but it helps the body to move food along the intestines. As the fibre moves through the large intestine, bacteria feed on it and together the fibre and bacteria add bulk to the food. This helps the muscles of the large intestine push the food along. Fibre also takes up water like a sponge and this makes the undigested foods, which form faeces, soft and easy to release from the body. If a person's diet lacks fibre, they may experience bowel problems such as constipation.

Vitamins

Vitamins are needed by the body in much smaller amounts than carbohydrates, fats and proteins. When vitamins were first discovered, they were named after letters of the alphabet. Later, when scientists knew more about them, they were also given chemical names.

Each vitamin has one (or more) use inside the body. Vitamin A is involved in allowing the eyes to see in dim light, and in making a mucus lining for the respiratory, digestive and excretory systems, which protects against infection from microorganisms.

A lack of vitamin C causes the **deficiency disease** called scurvy. As the disease develops, bleeding occurs at the gums in the mouth, under the skin and into the joints. Death may occur due to massive bleeding in the body.

Vitamin D helps the body take up calcium from food to make bones and teeth strong. Children who lack vitamin D in their diet may develop the deficiency disease called rickets, in which the bones do not develop to their full strength and may therefore bend. This is seen particularly in the leg bones, as shown in the X ray in Figure 4.5.

▼ **Table 4.1** Vitamins and their uses.

Vitamin	Effect on body	Good sources
A	increased resistance to disease helps eyes to see in the dark	milk, liver, cod-liver oil
C	prevents the disease scurvy in which gums bleed and the circulatory system is damaged	blackcurrant, orange, lemon, papaya, guava
D	prevents the disease rickets in which bones become soft and leg bones of children may bend	egg yolk, butter, cod-liver oil, pilchard, herring, sunlight

There are many stories about the discovery of vitamins. They show how scientists thought and worked scientifically to make their discoveries. Finding the cause of a disease called beriberi is one example.

Science in context

Thinking and working scientifically to discover a vitamin

Christiaan Eijkman (1858–1930) was a Dutch doctor who worked at a medical school in the East Indies in the late-nineteenth century. He investigated the disease called beriberi. In this disease, the nerves fail to work properly and the action of the muscles becomes weak. All movements, especially walking, become difficult and, as the disease progresses, the heart may stop. At this time, other scientists had recently shown that microorganisms caused a number of diseases. It seemed reasonable to think that beriberi was also caused by a microorganism of some kind. Eijkman set up some investigations to find out, but he was not having any success. Then one day a flock of chickens that were kept at the medical school began to show the symptoms of beriberi.

Eijkman tested them for signs of the microorganisms that he believed were causing the disease. Again, he had no success in linking the disease to the microorganisms but, while he was studying the chickens, they recovered from the disease.

▲ **Figure 4.6** A flock of hens with beriberi.

Eijkman began to search for a reason why they had developed the disease and also why they had recovered so quickly. He discovered that the chickens were usually fed on chicken feed (a specially prepared mixture of foods that keeps them healthy). A cook who had been working at the medical school had stopped using the chicken feed and had fed the chickens on rice that had been prepared for the patients. This cook had left and a new cook had been employed who would not let the rice be fed to the chickens. The birds were once again given chicken feed. When Eijkman fed the chickens rice again, they developed beriberi. When he fed them on chicken feed, they recovered from the disease straight away.

The rice fed to the chickens and the patients was polished rice. This meant it had its outer skin removed and appeared white. Later work by scientists showed that the skin of rice contains vitamin B1 (or thiamin). This vitamin is needed to keep nerves healthy and prevent beriberi.

Further scientific work on vitamin B1 showed it to be important in preventing digestive disorders. It is found in bread, milk, brown rice, soya beans and potatoes.

5 What are the symptoms of beriberi?

6 How serious is the disease?

7 Why did Eijkman begin by looking for microorganisms as a cause of beriberi?

8 In what way did chance play a part in the discovery of the cause of beriberi? Explain your answer.

9 How did Eijkman's work alter the way scientists thought diseases developed?

10 What is the danger in having a diet which consists mainly of polished rice? Explain your answer.

  11 Write down a plan for an investigation to check Eijkman's work on chickens and beriberi. How would you make sure it was fair and that the results were reliable?

> **LET'S TALK**
>
> Eijkman performed his experiments on animals. Question 11 asked you to plan an investigation to check his work. Your plan may have also featured studying animals. A great deal of information that benefits humans has been gathered by studying animals in experiments. Are there any guidelines that you would want scientists to follow in experiments involving animals?

12 A meal contains carbohydrate, fat, protein, vitamin D, calcium and iron. What happens to each of these substances in the body?

Minerals

The body needs 20 different **minerals** to keep it healthy. Two minerals that are important are calcium and iron. Calcium is needed to make strong bones and teeth. Iron is needed by the red blood cells, to help them take up oxygen in the lungs and transport it around the body to cells where it is needed. If a diet lacks iron, the red blood cells carry less oxygen around the body and a person can become tired and weak. This condition is called **anemia**, but it can be cured by increasing the amount of iron in the diet.

Water

About 70 per cent of the human body is water. The body can survive for only a few days without a drink of water.

All the substances that the body needs for life dissolve in water. Water is found in all cells and it is where all the chemical reactions between the substances take place to keep us alive.

The blood is made mainly from water. It is the liquid that transports all the other blood components around the body. Water is also used to cool down the body by the evaporation of sweat from the skin.

> **DID YOU KNOW?**
>
> Each mineral may have more than one use. For example, calcium is also used by the body to make your muscles work, and it helps your blood to clot when you cut yourself.

> **CHALLENGE YOURSELF**
>
> A watermelon can be over 90 per cent water. What experiment could you make to find out what percentage of a watermelon is water? If your teacher approves, try it.

The nutrients in food

The amounts of nutrients found in different foods have been worked out by experiment and calculation. The amounts are usually expressed for a sample of food weighing 100 g. Table 4.2 shows the nutrients in a small range of common foods.

▼ **Table 4.2** The nutrients in some common foods (the symbol 'μg' stands for 'microgram' and is a measure of a millionth of a gram).

Food/ 100 g	Protein/ g	Fat/ g	Carbohydrate/ g	Calcium/ mg	Iron/ mg	Vitamin C/ mg	Vitamin D/ μg
potato	2.1	0	18.0	8	0.7	8–30	0
carrot	0.7	0	5.4	48	0.6	6	0
bread	9.6	3.1	46.7	28	3.0	0	0
spaghetti	9.9	1.0	84.0	23	1.2	0	0
rice	6.2	1.0	86.8	4	0.4	0	0
lentil	23.8	0	53.2	39	7.6	0	0
pea	5.8	0	10.6	15	1.9	25	0
jam	0.5	0	69.2	18	1.2	10	0
peanut	28.1	49.0	8.6	61	2.0	0	0
lamb	15.9	30.2	0	7	1.3	0	0
milk	3.3	3.8	4.8	120	0.1	1	0.05
cheese 1	25.4	35.4	0	810	0.6	0	0.35
cheese 2	15.3	4.0	4.5	80	0.4	0	0.02
butter	0.5	81.0	0	15	0.2	0	1.25
chicken	20.8	6.7	0	11	1.5	0	0
egg	12.3	10.9	0	54	2.1	0	1.50
fish 1	17.4	0.7	0	16	0.3	0	0
fish 2	16.8	18.5	0	33	0.8	0	22.2
apple	0.3	0	12.0	4	0.3	5	0
banana	1.1	0	19.2	7	0.4	10	0
orange	0.8	0	8.5	41	0.3	50	0

Notes for Tables 4.2 and 4.3

Vegetables are raw; the bread is wholemeal bread; cheese 1 is cheddar cheese; cheese 2 is cottage cheese (a cheese with a milder (less strong) flavour than many other cheeses and a soupy texture); fish 1 is a white fish such as cod; fish 2 is an oily fish such as herring.

13 a Which foods would a vegetarian not eat?
   b Which foods would a vegetarian have to eat more of and why?

LET'S TALK

In groups, answer the following questions.
1 In Table 4.2, which foods contain the most
   a protein
   b fat
   c carbohydrate
   d calcium
   e iron
   f vitamin C
   g vitamin D?
2 a Which food provides all the nutrients needed?
   b Why might you expect this food to contain so many nutrients?

Can food labels be used to sort foods into nutrient groups?

You will need:

a selection of clean, empty packets and tins that contained a wide variety of foods. Each one should have a list of the nutrients in the food and the amounts that are present, as shown in Figure 4.7.

Nutrition		
Typical Composition	This pack (450g) provides	100g (3¹/₂oz) provide
Energy	2610kJ	580kJ
	621kcal	138kcal
Protein	13.2g	2.9g
Carbohydrate	82.3g	18.3g
of which sugars	18.0g	4.0g
Fat	26.6g	5.9g
of which saturates	13.5g	3.0g
mono-unsaturates	10.4g	2.3g
polyunsaturates	2.7g	0.6g
Fibre	7.2g	1.6g
Sodium	1.8g	0.4g
A serving (450g) contains the equivalent of approx. 4.5g of salt.		

▲ **Figure 4.7** The nutrients in a food product are displayed on the side of the packet.

In this enquiry, just look at the total amounts of protein, fat and carbohydrate in 100 g of each food.

Investigation and recording data

1　Arrange the food labels in order of amount of protein, starting with the food containing the most protein and ending with the food containing the least. Classify the foods into high-protein foods, low-proteins foods and a group which has amounts of protein that are neither very high or low, which you could call the average protein group.

2　Repeat step 1 with fats.

3　Repeat step 1 with carbohydrates.

4　Present your data in the form of a table.

Energy in food

A piece of equipment called a bomb **calorimeter** is used to find out how much energy there is in food. A food item is placed inside it and oxygen is pumped in, then the food is set on fire and all the heat that is produced is measured. Finally, the energy in 100 g of food is calculated in kilojoules (kJ) and tables are produced like the ones shown in Table 4.3.

▼ **Table 4.3** The energy value of some common foods.

Food/100 g	Energy/kJ
potato	324
carrot	98
bread	1025
spaghetti	1549
rice	1531
lentil	1256
pea	273

Food/100 g	Energy/kJ
jam	1116
peanut	2428
lamb	1388
milk	274
cheese 1	1708
cheese 2	480
butter	3006

Food/100 g	Energy/kJ
chicken	602
egg	612
fish 1	321
fish 2	970
apple	197
banana	326
orange	150

14 Table 4.3 shows the amount of energy provided by 100 g of each of the foods shown in Table 4.2.
   a Arrange the nine foods with the highest energy values in order, starting with the highest and ending with the lowest.
   b Look at the nutrient content of these foods in Table 4.2. Arrange the nine foods from question part **a** into groups according to whether you think the energy is stored as fat or as carbohydrate.
   c Do fats and carbohydrates store the same amount of energy (see also pages 30–31)? Explain your answer.

15 Why might people who are trying to lose weight eat cottage cheese instead of cheddar cheese?

16 Mackerel is an oily fish. Describe the nutrients you would expect it to contain.

17 Look again at the eating pattern you prepared for Question 1 on page 26. Analyse your diet and divide it up into the food groups shown in Table 4.5 (page 36). How well does your diet provide you with all the nutrients you need?

Can food labels be used to sort foods into energy groups?

You will need:

the selection of clean, empty packets and tins that contained a wide variety of foods that you used in the previous enquiry.

In this enquiry, just look at the total amount of energy in kilojoules in 100 g of each food.

Plan, investigation and recording data

1 Arrange the food labels in order, starting with the food that has the highest energy value and ending with the food that has the least. Classify the foods into high-energy foods, low-energy foods and a group which has energy values that are neither very high or low, which you could call the average energy group.
2 Present your data in a table.

Examining the results

1 Compare your table with the one you made in the previous enquiry and answer the following questions.
2 Can you see a pattern or trend between the amount of energy in a food and
   a the amount of protein
   b the amount of fat
   c the amount of carbohydrate?

Explain your answers.

Conclusion

Do the results of the investigation support the question of the enquiry? Draw a conclusion and explain your answer.

▼ **Table 4.4** Average daily energy used by males and females.

| Age/years | Daily energy used/kJ | |
	Males	Females
2	5500	5500
5	7000	7000
8	8800	8800
11	10000	9200
14	12500	10500
18	14200	9600
25	12100	8800

18 Table 4.4 shows how the energy used by an average male person and an average female person changes between the ages of 2 and 25 years.
  a Plot the information given in the table as a single graph.
  b Describe what the graph shows.

19 a Explain why there is a difference between the amounts of energy used by a 2-year-old child and an 8-year-old child.
  b Explain why there is a difference between the amounts of energy used by an 18-year-old male and an 18-year-old female.
  c Explain why there is a change in the amount of energy used as a person ages from 18 to 25.

20 What changes would you expect to see in the amount of energy used by:
  a a 25-year-old person who changed from a job delivering letters and parcels to working with a computer
  b a 25-year-old person who gave up working with computers and took a job on a building site that involved carrying heavy loads
  c a 25-year-old female during pregnancy?

Keeping a balance

To remain healthy, a diet has to be balanced with a body's needs. A **balanced diet** is one in which all the nutrients are present in the correct amounts to keep the body healthy. You do not need to know the exact amounts of nutrients in each food to work out whether you have a healthy diet. A simple way is to look at a chart showing food divided into groups, with the main nutrients of each group displayed (see Table 4.5). You can then see if you eat at least one portion from each group each day and more portions of the food groups that lack fat. Remember that you also need to include fibre even though it is not digested. It is essential for the efficient movement of food through the large intestine. Fibre is found in cereals, vegetables and pulses, such as peas and beans.

▼ **Table 4.5** The groups of foods and their nutrients.

Vegetables and fruit	Cereals	Pulses	Meat and eggs	Milk products
carbohydrate	carbohydrate	carbohydrate	protein	protein
vitamin A	protein	protein	fat	fat
vitamin C	B vitamins	B vitamins	B vitamins	vitamin A
minerals	minerals	iron	iron	B vitamins
fibre	fibre	fibre		vitamin C
				calcium

21 Look again at the food labels from the enquiry. Of these, choose the labels of the foods you regularly eat, and look at the amounts of protein, fat and carbohydrate listed on them. Do they show evidence of a balanced diet? Explain your answer.

22 Now look at the amount of energy on the food labels from the enquiry. Again, look at the foods that you regularly eat. Do they show evidence of a high-energy diet or a low-energy diet? Explain your answer.

Science extra: Treating iodine deficiency

Iodine is a mineral that the body needs in order to repair its damaged cells, to control its growth and to release energy from food. If it is lacking in the diet, it causes a swelling in the neck called **goiter**, changes in the beating of the heart and feelings of tiredness and dizziness.

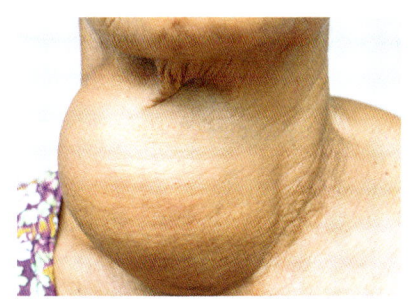

▲ **Figure 4.8** The neck of a person with goiter.

Iodine passes into the roots of plants from the soil, but in places where the iodine has been washed out of the soil over a long time, there is little iodine that the plants can take up. This results in a lack of iodine in food.

One place where body changes due to iodine deficiency have been observed amongst the population is in Tanzania, in Africa. Scientists responded to this by giving people salt that contained iodine through a process called universal salt iodation (USI). This produced an increase in iodine in the diet and improved the health of the people living there.

CHALLENGE YOURSELF

Look at the homepage of the Iodine Global Network (www.ign.org) and find the world map. Click on your country to see if the population's iodine intake is adequate or insufficient. Survey the map to find at least six countries where the intake is insufficient.

---

23 Do you see a pattern in the countries where iodine intake is insufficient, according to the Iodine Global Network? For example, is the country an island, or in the middle of continent, or does it have a coast-line? Explain your answer.

LET'S TALK

Is it right to add nutrients to people's diets in addition to the ones they get from eating their own food? Explain your answer.

LET'S TALK

Do you snack between meals? If so, are they high-energy snacks? What steps would you take to reduce their damage to your balanced diet?

DID YOU KNOW?

Cooking helps us digest food and absorb more nutrients from it. However, some cooking methods, such as steaming or stir-frying, help release more nutrients than other methods, such as boiling.

24 What does the food pyramid tell you about eating
   a high-carbohydrate foods such as bread, pasta and rice
   b foods rich in vitamins, minerals and fibre, such as fruit and vegetables
   c protein-rich foods, such as milk products, meat and fish
   d foods containing a large amount of fat, such as cheese and chocolate?

A healthy diet

We have seen that the body needs a range of nutrients to keep it healthy, and this can be provided by a balanced diet. Everyone should aim to eat a balanced diet, but diets can be unbalanced easily and quickly by some eating habits, such as regularly eating high-energy snacks like sweets, chocolate, crisps and ice cream between meals. These habits can lead to a number of unhealthy conditions such as obesity (see page 43) and tooth decay.

There are alternatives to high-energy snacks which, in addition to having less energy, also have more vitamins and minerals. Fruits and raw vegetables, such as celery, tomatoes and carrots, are example of these alternative snacks.

There is a lot of information about balanced diets, but an easy way to think about it is by looking at the pyramid of food in Figure 4.9.

It shows that you can eat large amounts of the foods shown at the base, but should eat smaller amounts of the foods in the sections moving up the pyramid, and only eat very small amounts of foods shown at the top.

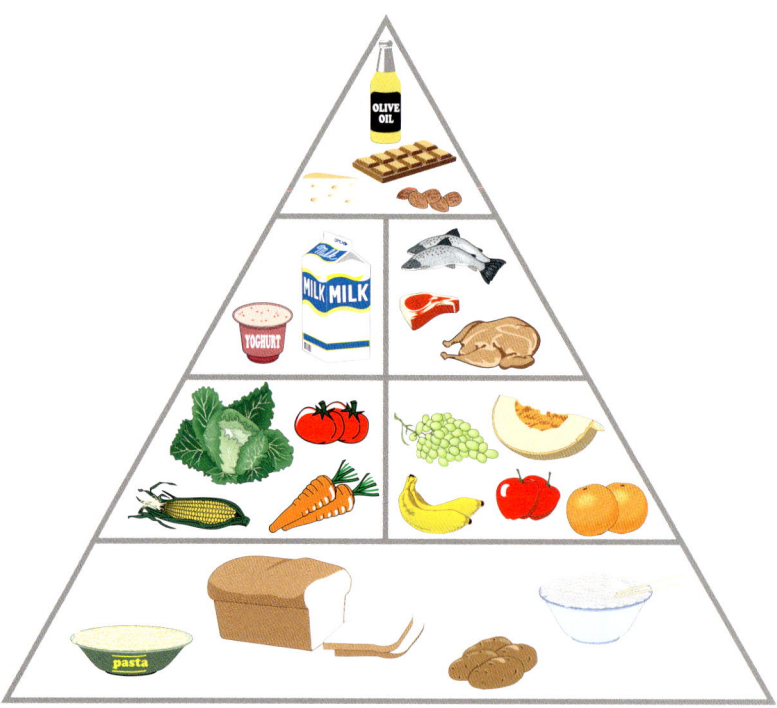

▲ **Figure 4.9** A pyramid of food.

25 Make a pyramid of food representing your diet. Describe how it compares with the pyramid in Figure 4.9. If it does not match the pyramid in the diagram, how can you change your diet to make it healthier?

Treating malnutrition and famine

Malnutrition occurs if a diet provides too few or too many nutrients. For example, anemia is caused by having too little iron in the diet, while obesity (see page 43) can be caused by having too much fat in the diet.

Famine is a lack of all foods, which can be caused by crop failure, but also by human activities such as war, where people have to move away from their crops and source of food. The lack of food affects the health of everybody, but it greatly affects the growth and development of young people. If there is just enough food for the young people to survive, their bodies will grow and develop much more slowly than normal and, as time goes on, their bodies will be smaller and less well-developed than people with healthy diets.

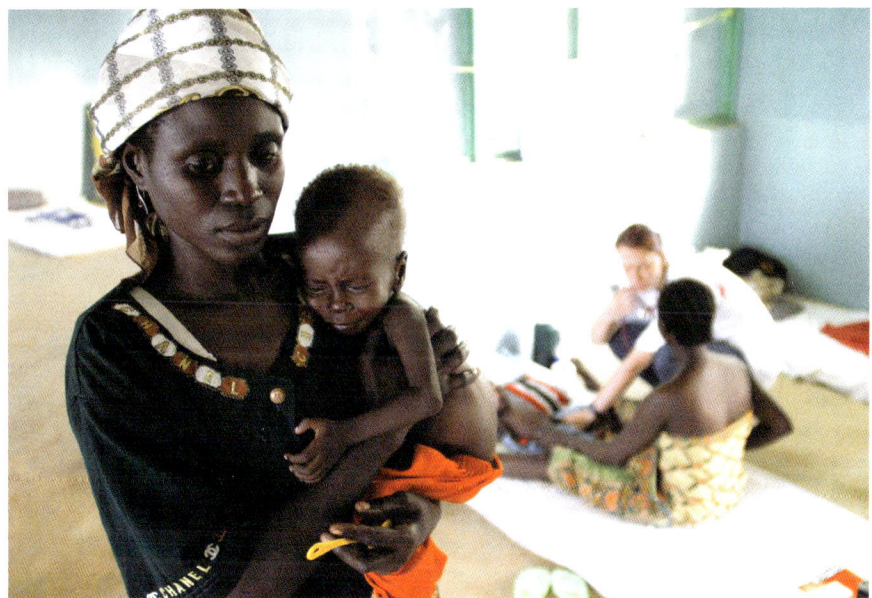

▲ **Figure 4.10** This mother has not had access to a balanced diet and so her baby is small.

In areas of famine, aid workers give malnourished children powdered milk mixed with water. There are two forms of powdered milk, called F75 and F100. Children are first given F75 which prepares their body to start digesting food again. Then, after a few days, they are fed F100. This contains more proteins and energy to help the body begin to build up again. The children may also be given an alternative to F100, called Plumpy'nut, which is made from peanut paste, powdered milk, minerals, oil, sugar and vitamins. A second food called Unimix is also added to the diet. This is a flour made from maize and soya beans, to which powdered milk, minerals, oil, sugar and vitamins are added. It is mixed with water to make a porridge. The children may be fed on this diet for up to a month as their bodies recover, and they can then move on to other foods.

26 a Look at Table 4.2 on page 32 and compare the nutrients in rice and peanuts.

b Why do you think a food based on peanuts is better than one based on rice in order to help people recovering from **starvation** and malnutrition?

▲ **Figure 4.11** Children receiving food aid at a distribution centre in Somalia.

Summary

- ✔ A balanced diet is made up of proteins, carbohydrates, fats and oils, water, minerals and vitamins in the right quantities.
- ✔ Nutrients are needed by the body to keep it healthy and can be split up into carbohydrates, vitamins, fats and oils, minerals and proteins.
- ✔ Carbohydrates and fats can be used as an energy store in animals.
- ✔ Animals consume food to obtain energy and nutrients.
- ✔ Science in context: Christian Eijkman's work with chickens and beriberi led to the discovery of a vitamin (B1).
- ✔ We can work out the amount of different nutrients in food as well as the energy it contains and use this information to make healthy food choices.
- ✔ Science in context: We can treat malnutrition using foods such as powdered milk.

End of chapter questions

1 What does the body need the following for?

   a glucose

   b fats

   c proteins

2 Why does the body need

   a calcium

   b iron?

3 What foods can people eat to prevent them getting scurvy?

4 What foods should people eat to prevent them getting rickets?

5 Imagine you knew of someone who ate a lot of high-energy snacks.

   a What could happen to them if they continued with this habit?

   b What advice would you give them to keep them healthy?

6 **a** What is a balanced diet?

   b Give examples of the quantities of different foods you would eat to keep your diet balanced.

 Now you have completed Chapter 4, you may like to try the Chapter 4 online knowledge test if you are using the Boost eBook.

5 | A healthy lifestyle

In this chapter you will learn:
- how your health, growth and development can be affected by how you live, what you eat and habits such as smoking
- about the role of dietitians (Science in context)
- about cholesterol and the circulatory system (Science in context).

Do you remember?

- Which organ pumps the blood around the body?
- Which organs move your bones when you exercise?
- Name the five nutrient groups you need in your diet.
- What does the word malnutrition mean?
- Why do you need healthy lungs?

Your lifestyle is the way in which you live. A healthy lifestyle is the way in which you live to keep your body as healthy as possible. The key features of a lifestyle to encourage your healthy growth and development are diet, keeping a healthy heart, taking exercise and avoiding smoking.

Diet

1 How does the pyramid of food help people to plan a balanced diet?

In Chapter 4, diet was examined in detail, and a pyramid of food (Figure 4.9 on page 38) was shown to help you construct a balanced diet. Everyone needs a balanced diet from birth for healthy growth and development. Here are some ways in which an unbalanced diet can affect human growth and development.

Malnutrition

Malnutrition and famine were featured in Chapter 4, but malnutrition can also occur in people who are not in famine-affected areas.

Deficiency diseases

When a lack of a vitamin or mineral occurs, it can produce a deficiency disease. We saw on page 29 that a lack of vitamin C produces the deficiency disease called scurvy and a lack of vitamin D causes rickets. When there is a lack of iron in the diet, the deficiency disease called anemia occurs. Iron is needed to help red blood cells transport oxygen around the body. If there is a lack of iron, there is a lack of haemoglobin and therefore a lack of oxygen, so the person with this disease feels permanently tired.

Anorexia nervosa

In the condition anorexia nervosa, a person eats very little and fears weight gain. This results in too little high-energy food being eaten and the body becomes thin because it uses up energy stored as fat. Energy stored as protein may also be used up. As the body's stored energy is depleted, the body undergoes extreme weight loss which can result in death. Anorexia nervosa occurs mainly in teenage girls, but is occurring increasingly in teenage boys and adult men and women. As soon as the condition is diagnosed, the patient needs careful counselling to give them the best chance of making a full recovery.

Obesity

Obesity is a condition which develops in children (and adults) who have a very high-energy diet. The diet may be made up of lots of sweets, chocolate, crisps, pizza or other foods with a high carbohydrate and fat content. The energy is stored in the body as fat, and the increased body mass makes the person tend to move around less than normal.

As the person uses less energy than before but continues to eat the high-energy diet, even more body fat is produced. This can affect the working of the heart (see page 45) and the increase in body mass puts extra strain on their bones and joints. Obesity can lead to a form of **diabetes** called type 2 diabetes. This affects the way that the body controls the amount of digested sugar in the blood. Normally, the body produces a substance called **insulin** which helps the body to take extra sugar out of the blood and store it in the liver. In type 2 diabetes, the body no longer responds to the insulin, and blood with a high sugar content flows around the body. This can eventually cause blindness and damage to the nerves and blood vessels in the feet, which can only be treated by amputation. It can also cause kidney damage which can be fatal.

▲ **Figure 5.1** This is an example of a diet containing too much energy.

2 How do obese people put their health and lives at risk?

Treating obesity

As obesity develops due to an unbalanced diet, one part of the treatment is to help the person change to a balanced diet. The second part is to build up an exercise programme so that more energy is used up and the body is strengthened. In the treatment of obese adults, the aim is to lose body mass, but the treatment of obese children is slightly different. This is because children are still growing. The aim in the treatment of obese children is to help their bodies grow into the correct proportions so that over a few years the child's body mass and appearance become more normal.

3   Could you be a dietitian? Explain your answer.

Science in context

Dietitians

A dietitian is a person who has studied the science of food and nutrition to a high level and uses this knowledge to help others. Dietitians usually work in hospitals, health centres and clinics, where they advise people about diet and lifestyle in order to make improvements in their health. The people they help can have a wide range of conditions, from diabetes and food allergies, to eating disorders and kidney failure.

Many countries across the world have dietitians who are working to improve the health of their people.

▲ **Figure 5.2** A dietitian.

A healthy heart

The heart may beat up to 2 500 million times during a person's life. Its function is to push blood around the 100 000 km of blood vessels in the body. This push creates a **blood pressure** that drives the blood through the blood vessels. As the parts of the heart fill with blood, the pressure in the blood vessels is reduced, but as the heart pumps the blood out along the arteries, the blood pressure rises. (Pressure is produced when a force acts over an area, such as pushing a drawing pin into a board. When it is used in the term 'blood pressure' it describes the force used by the heart to pump blood around the body.)

The walls of arteries are elastic so they stretch and contract with the blood pressure. In young people, the arteries are free from obstructions and their diameters are large enough to let the blood flow with ease. As the body ages, the **artery** walls become less elastic.

The heart has its own blood vessels called the coronary arteries and **veins**, which transport blood to and from the heart muscle. The coronary blood vessels are shown in Figure 5.3.

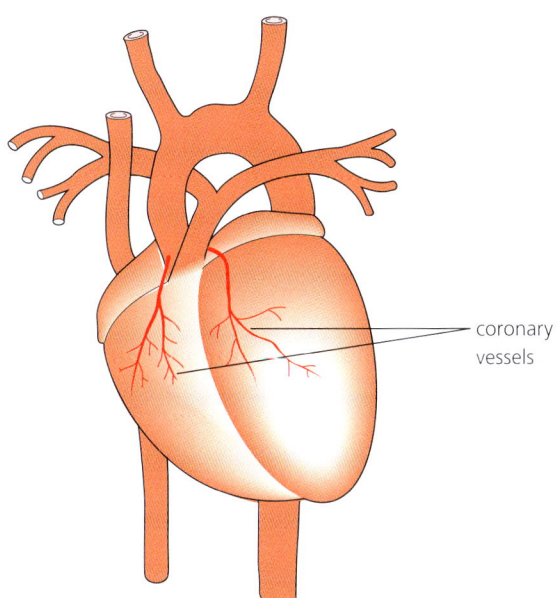
coronary vessels

▲ **Figure 5.3** The coronary blood vessels.

Over time, fatty substances, such as **cholesterol**, stick to the walls of an artery. Calcium settles in the fatty layer and forms a raised patch on the inside wall of the artery. The blood then has less space to pass along the artery and its pressure rises as it pushes through the narrower tube. Other components of the blood, such as platelets, settle on the patch and make it larger. This may cause a **blood clot** which narrows the artery even more or can completely block it, causing a **thrombosis**. This means that the artery is unable to supply oxygen and other nutrients to the organ that the artery is supplying with blood. A thrombosis in a coronary artery causes a heart attack. A thrombosis in an artery in the **brain** causes a stroke.

DID YOU KNOW?
Sometimes the word 'coronary' is used to describe a heart attack. This event is also known as a cardiac arrest.

Science in context

Cholesterol and the circulatory system

In the early twentieth century, Russian scientist Nikolay Anichkov (1885–1964) was studying factors which affect health. He tested rabbits with a diet high in cholesterol and found that it caused patches to form inside arteries. Other scientists at the time did not see any future in this work as it was about rabbits.

About forty years later, American scientist John Gofman (1918–2007) had developed an interest in trying to link cholesterol in human blood to the patches that developed inside arteries. He believed that Anichkov's work provided evidence that supported his ideas. He began experiments to test his ideas, but again other scientists ignored the work for many years.

In time, American scientist Ancel Keys (1904–2004) studied the work that had been done on cholesterol, and developed a hypothesis that high levels of cholesterol in the diet were responsible for the patches in the arteries and the diseases they caused. He set up an enquiry called the Seven Countries Study

▲ **Figure 5.4** Nikolay Anichkov.

4 What stopped other scientists from following up Anichkov's work with more experiments?

5 If you were a scientist from the past, how do you think Gofman's work would be of more interest to you in studying human diseases than Anichkov's work?

6 Why do you think it was a good idea to study people living in different countries across the world?

(SCS) in which scientists investigated the lifestyles, diets and diseases of people in the USA, Finland, the Netherlands, Italy, Yugoslavia (now Serbia and Croatia), Greece and Japan. The data collected in this enquiry supported his hypothesis and led to a review of diets to prevent heart attacks and strokes.

Cholesterol is needed to keep the body healthy, but if it is present in a high concentration in the blood it can damage the circulatory system. Many studies have been made on the causes of high cholesterol and advice has been produced in order to help people to keep their cholesterol levels healthy. These include:

- reducing the intake of substances called trans fats which are used in the making of processed food, such as biscuits, snack foods, and 'fast foods'
- eating fresh vegetables and fruit
- eating lean meats (with small amounts of fat in them) and poultry, such as chicken, and fish
- eating beans, nuts and whole grains, such as brown rice, barley and quinoa
- choosing non-fat or low-fat dairy products, such as low-fat milk and cheese.

▲ **Figure 5.5** John Gofman.

▲ **Figure 5.6** Ancel Keys.

CHALLENGE YOURSELF

Studies continue on the links between diet and disease. Use the internet to find out the latest advice about diet and keeping cholesterol at a healthy level in the blood.

How do your findings compare with the results of earlier studies?

The features that develop in the body that cause heart disease can be inherited. People whose relatives have had heart disease should take special care to keep their heart and circulatory system healthy.

You can find out more about the action of the heart by taking your **pulse**.

CHALLENGE YOURSELF

Measuring your pulse

You learnt how to measure your pulse rate last year. Here is a reminder.

Process

1 Hold out your right hand with the palm up.
2 Put the thumb of your left hand under your wrist.
3 Let the first two fingers of your left hand rest on the top of your wrist.
4 Feel around on your wrist with these two fingers to find a throbbing artery. This is your pulse.
5 You can measure your pulse rate by counting how many times your pulse beats in a minute.

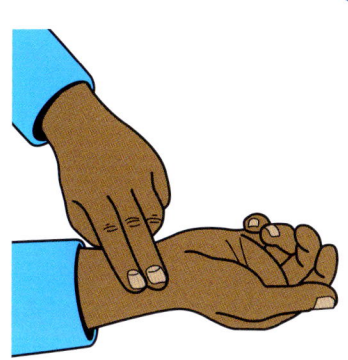

▲ **Figure 5.7** Measuring a pulse.

Some people find the pulse beats per minute by taking a pulse for 15 seconds, then multiplying the number of beats by 4. You can check the accuracy of this approach in the following enquiry.

Can you take a pulse accurately?

You will need:

a stop-clock or timer.

Plan, investigation and recording data

1 Take your pulse while sitting down for 1 minute and record the beats per minute.
2 Take your pulse while sitting down for 15 seconds then multiply the number of beats by 4 and record the beats per minute.
3 Construct a plan to make the results more reliable. If your teacher approves, try it.
4 Construct a plan in which someone else takes your pulse to test your results. If your teacher approves, try it.

Examining the results

Compare the data from your own measurements and from the measurements of someone else.

Conclusion

Draw a conclusion. Which method is the most accurate, or are they equally accurate?

Is your conclusion limited in some way? Explain your answer.

What improvements could be made? Explain the changes that you suggest.

A healthy heart is needed to transport oxygen and digested food in the blood to all the cells of the body to keep them alive and working correctly. Wastes are also transported away.

7 Why do you need a healthy heart for healthy growth and development? Explain your answer.

Exercise and the heart

Activity and the heart

The heart is made of muscle and, like all muscles, it needs exercise if it is to remain strong. The heart muscles are exercised when you take part in the activities like those listed in Table 5.1 (page 48). The heart muscle contracts more quickly and more powerfully during exercise than it does at rest so that more blood can be pumped to your muscles. These muscles need more blood to provide them with extra oxygen while they work.

Regular exercise makes many of the **organ systems** become more efficient. It also uses up energy and helps to prevent large amounts of fat building up in the body.

Exercise can increase your fitness in three ways: it can improve your strength, it can make your body more flexible and less likely to experience sprains, and it can increase your endurance, which is your ability to exercise steadily for long periods without resting. Different activities require different levels of fitness. Table 5.1 shows these levels for different sporting activities. By studying the table you can work out which activities you could do to develop one or more of the three components of fitness.

8 How do you think the effect of exercise on the body's organ systems helps in growth and development?

9 Which activities demand great flexibility?

10 Which activity is the least demanding?

11 Which activities are the most demanding?

12 How do the demands of soccer and long-distance running compare?

13 Which activity would you choose to do from Table 5.1? What are its strengths and weaknesses?

▼ **Table 5.1** Levels of fitness required for different activities.

Activity	Strength	Flexibility	Endurance
basketball	✓✓	✓✓	✓✓✓
dancing	✓✓	✓✓✓	✓✓
golf	✓✓	✓✓	✓✓
long-distance running	✓✓✓	✓✓	✓✓✓
soccer	✓✓	✓✓	✓✓✓
rugby	✓✓✓	✓✓	✓✓✓
squash	✓✓✓	✓✓✓	✓✓✓
swimming	✓✓✓	✓✓	✓✓✓
tennis	✓✓✓	✓✓✓	✓✓✓
walking	✓	✓	✓✓

◀ **Figure 5.8** Playing team sports is a great form of exercise.

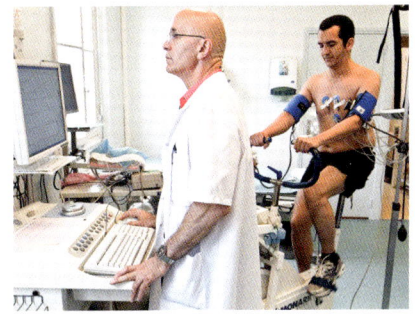

▲ **Figure 5.9** Pedalling an exercise bicycle makes the heart beat faster in order to provide blood for the leg muscles.

Care with exercise

When people decide to get fit, they may choose one of the activities from Table 5.1 and begin with great enthusiasm. However, they may experience a sprain or pains by trying to exercise too hard too early.

The skeleton and muscles work together to provide movement. Someone who has not been active for a long time may need to build up their exercises gradually so that the muscles and joints can become adapted to the increased activity. If this is not done and someone receives an injury early on in their exercise programme, they may decide not to continue and they will become unfit again. It can help to be aware of how the skeleton and muscles work and to think of them when an exercise programme is begun.

14 Many people claim that they do not have time to exercise. How would you motivate such people to take some form of exercise? Which activities might suit them best?

In Chapter 3 you investigated breathing rate when at rest and after exercise. Find out if there is a link between pulse rate (and heartbeat) and exercise in the following enquiry.

15 What did you conclude from your previous enquiry into breathing rate and exercise?

Is there a link between pulse rate and exercise?

You will need:

a stop-clock or timer and an open space for exercise.

Hypothesis

The breathing rate changed after exercise, so the pulse rate (and heart rate) may also change after exercise.

Prediction

Make a prediction based on the hypothesis.

Plan and investigation

Make a plan to test the hypothesis and prediction and, if your teacher approves, try it.

Examining the results

Examine the data you have collected. Do you have enough evidence to support your conclusion? Explain your answer.

Conclusion

Draw a conclusion after comparing your evaluation with the hypothesis and prediction. Was your hypothesis testable? Explain your answer.

Smoking and health

We have seen how the respiratory system works to allow the exchange of respiratory gases. An efficient exchange is needed for good health. When people smoke they damage their respiratory system and risk seriously damaging their health.

There are over a thousand different chemicals in cigarette smoke, including the highly addictive nicotine. These chemicals swirl around the air passages and touch the air passage linings when a smoker inhales. In a healthy person, dust particles are trapped in mucus and move up to the throat by the beating of microscopic hairs. These small amounts of dust and mucus are then swallowed. In a smoker's respiratory system, the microscopic hairs stop beating because of chemical damage done by the smoke. More mucus is produced, but instead of being carried up by the cilia it is coughed up by a jet of air as the smoker exhales strongly. This is known as 'smoker's cough' and the amount of dirty mucus reaching the throat may be too much to swallow.

In time, chronic bronchitis may develop. This is when the lining of the bronchi becomes **inflamed** and open to infection from microorganisms. The inflammation of the air passages makes breathing more difficult, and the smoker develops a permanent cough. The coughing causes the walls of some of the alveoli in the lungs to burst. When this happens, the surface area of the lungs in contact with the air is reduced. This leads to a disease called **emphysema**.

16 What is the function of a smoker's cough?

17 Why may chronic bronchitis lead to other diseases?

18 How does a reduced number of alveoli affect the exchange of oxygen and carbon dioxide?

19 Someone with emphysema breathes more rapidly than a healthy person. Why do you think this is?

LET'S TALK

Should a person who becomes ill through having an unhealthy lifestyle receive the same amount of medical attention as someone who has had an accident?

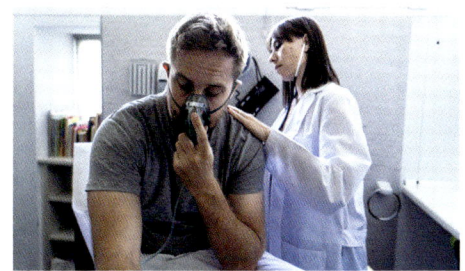

▲ **Figure 5.10** People with lung damage from smoking need extra oxygen and regular checks on their lungs.

Some of the cells lining the air passages are killed by the chemicals in the smoke. They are replaced by cells below them as they divide and grow. Some of these cells may be damaged by the smoke too, and as they divide, they may form cancer cells. These cells replace the normal cells in the tissues around them but they do not perform the functions of the cells they replace. The cancer cells continue to divide rapidly and form a lump called a **tumour**. This may block the airway or break up and spread to other parts of the lung where more tumours can develop.

Smoking and growth and development

We have seen how smoking can damage the lungs but it can also affect the growth and development of the body. The damage to the lungs prevents them from working efficiently and supplying oxygen in the correct quantities to the body cells.

Pregnant women who smoke risk reducing the oxygen supply to their developing babies. This results in the babies being born at a lower than normal birth weight which could affect further development as the child grows up and may also lead to heart disease and diabetes.

Adults who smoke may experience a loss of appetite which means they eat less food and reduce the amount of nutrients they need. This causes them to lose body weight and may affect their general health.

Helping smokers stop

20 How are cancer cells different from normal cells in the lung tissue?

21 Why do cancer cells in an organ make the organ less efficient?

22 Why might the growth of cancer tumours in an organ have fatal results?

People smoke tobacco because they become addicted to the drug nicotine which is one of the chemicals in the tobacco leaf. An **addiction** is a condition in which a person is unable to live a normal life without smoking, drinking alcohol or taking drugs. Smokers who wish to give up and break their addiction can receive advice and support from health services. They may chew nicotine gum or attach patches containing nicotine to their skin so that they still receive some nicotine while not smoking. This helps them to break the habit of lighting a cigarette. It is claimed that e-cigarettes or vapes can help people to stop smoking, but some research suggests they are just as dangerous as smoking cigarettes, and they have been banned in some countries.

Summary

✔ One part of a healthy lifestyle is having a balanced diet. An unbalanced diet can lead to malnutrition, deficiency diseases, anorexia nervosa or obesity.

✔ A healthy heart is key to a healthy lifestyle. Poor blood pressure or high cholesterol can lead to heart conditions and illnesses, such as blood clots, thrombosis or a stroke.

✔ Regular exercise increases the efficiency of many of the organs, such as the heart and lungs, as well as decreasing the chance of injuries and illness.

✔ Smoking affects health, growth and development and can cause illnesses such as cancer, emphysema and 'smoker's cough'.

End of chapter questions

1 What can happen in an artery if there is too much cholesterol in the diet?

2 What is a thrombosis?

3 What causes a heart attack?

4 What causes a stroke?

5 What should you include in your diet to stop you getting a high level of cholesterol?

6 What disease could you develop if your diet lacks
 a vitamin C
 b vitamin D
 c iron?

7 a How does smoking affect the lungs?
 b What diseases could a smoker develop?

8 Four people took their pulse (measured in beats per minute, on a portable heart monitor) at rest, straight after exercise, 1 minute after exercise, 2 minutes after exercise and 3 minutes after exercise. Here are their results:

Anwar: 71, 110, 90, 79, 71

Baylee: 74, 115, 89, 77, 73

Chiumbo: 73, 125, 115, 108, 91

Daisy: 53, 80, 71, 46, 84

 a Make a table of the results.
 b Plot a graph of the results.
 c What trend can you see in the results?
 d When did Anwar and Baylee's hearts beat at the same rate?
 e Anwar claims to be fitter than Chiumbo. Do you think the results support his claim? Explain your answer.
 f Which result does not follow the trend? Explain why this may be so.

CHALLENGE YOURSELF

How long does it take your heart to return to its resting heartbeat after running for 2 minutes? Plan an investigation to find out and, if your teacher approves, try it.

9 A survey was made to find out about the smoking habits of young people aged 11–14. For each year group, 1000 boys and 1000 girls were asked if they were occasional smokers or regular smokers. The results are shown in Table 5.2.

▼ **Table 5.2**

Smokers	11 years	12 years	13 years	14 years
boys, occasional smokers	20	40	70	80
boys, regular smokers	10	30	60	140
girls, occasional smokers	15	0	90	110
girls, regular smokers	10	35	70	160

a How is the data for occasional and regular smokers for boys similar?

b Where is there an anomalous result? Give an explanation for it.

c What is the difference in the number of regular smokers between boys and girls at the age of 14?

d What percentage of the boys aged 12 are regular smokers?

e Do you predict the percentage of boys and girls who are regular smokers will go up, stay the same or go down for boys and girls aged 15?

f Which group, boys or girls, do you think will have the larger percentage of occasional smokers at the age of 15?

Now you have completed Chapter 5, you may like to try the Chapter 5 online knowledge test if you are using the Boost eBook.

In this chapter you will learn:

- about exploring ecosystems (Science in context)
- about the different ecosystems on the Earth
- about a variety of habitats that exist within an ecosystem
- about the impact of the bioaccumulation of toxic substances on an ecosystem
- about biomagnification (Science extra)
- what happens when there is a poison in the food chain (Science in context)
- how a new species and invasive species can affect other organisms and an ecosystem.

Do you remember?

- What is a food chain?
- Give an example of a food chain.
- What do these terms mean: producer, consumer, herbivore, omnivore, carnivore, predator and prey?
- What is a habitat?
- What does the term 'environment' mean?
- What is the difference between a food chain and a food web?
- Are microorganisms important in a habitat? Explain your answer.

▲ **Figure 6.1** Even a city street is part of an ecosystem.

The ecosystem you live in

An ecosystem is a description of how a community of plants, animals and microorganisms living in a particular place react together and with physical features such as the rocks, soils and weather conditions found in that place.

You are living in an ecosystem. Look around you. The chances are you may see some plants. A bird may fly across the sky or an insect may buzz as it passes by. Even on a city street you may see these things, as they form part of a city street ecosystem.

Most ecosystems have more components than this. If you move to a park or the countryside you will see many more plants, birds and insects, and members of other animal groups too, such as mammals, amphibians and molluscs.

CHALLENGE YOURSELF

Look out of the window and describe the ecosystem you can see. If you try this in class, keep your notes secret and then share them with others who have done this challenge. Make a class description of the ecosystem around your school.

DID YOU KNOW?

One of the oldest ecosystems on the planet is the Daintree Forest in Australia. It is estimated to be about 189 million years old!

▲ **Figure 6.2** Ecosystems are more diverse in greener, less urban environments.

All these living things react with the non-living parts of the ecosystem. In the street the non-living parts are the glass and concrete walls, the tarmac road and the hot air containing exhaust gases from all the vehicles moving along. In the park, the non-living parts are the soil under the grass (the minerals it contains and particles which let the water drain through it) and the cleaner air with its greater range of temperature due to the weather.

Science in context

Exploring ecosystems

About 250 years ago, explorers cut their way through rainforests, trekked across plains and climbed mountains to gather data about our world.

Alexander von Humboldt (1769–1859), a German explorer, made many journeys, and from the data he collected, he noticed that the number of species living in a place increased as he moved from the north to the tropics. The idea of different worldwide habitats began to form.

Nain Singh Rawat (1830–1882) was an Indian explorer who surveyed the Himalayas, known as the 'roof of the world', and a large length

▲ **Figure 6.3** Alexander von Humboldt explored many new habitats in South America.

DID YOU KNOW?
Huge ecosystems which cover the land areas of the planet are called **biomes**.

1 Look at a map of the world (Figure 6.5) and find your country. In which biogeographical realm do you live?

2 What pattern did von Humboldt see in his data?

3 What pattern did Sclater see in the data he studied?

4 What evidence from a secondary source and evidence from first-hand experience did Wallace use to help the development of the map shown in Figure 6.5?

5 What was the idea that started Möbius' investigations?

of the Bramaputra river. His data was recognised as very valuable by the Royal Geographical Society in London and the Paris Geographical Society.

Philip Sclater (1829–1913) was an English biologist who looked at the distribution of songbirds, such as the sparrow, around the world and decided that the Earth could be divided up into regions based on the birds found there. These regions are known as biogeographical realms (huge areas of land and oceans).

Alfred Russel Wallace (1823–1913), a Welsh biologist who travelled widely, found that other animals he had seen also fitted into Sclater's system of realms. The system continued to be developed, and a modified version of it is still in use today (Figure 6.5).

▲ **Figure 6.4** Nain Singh Rawat.

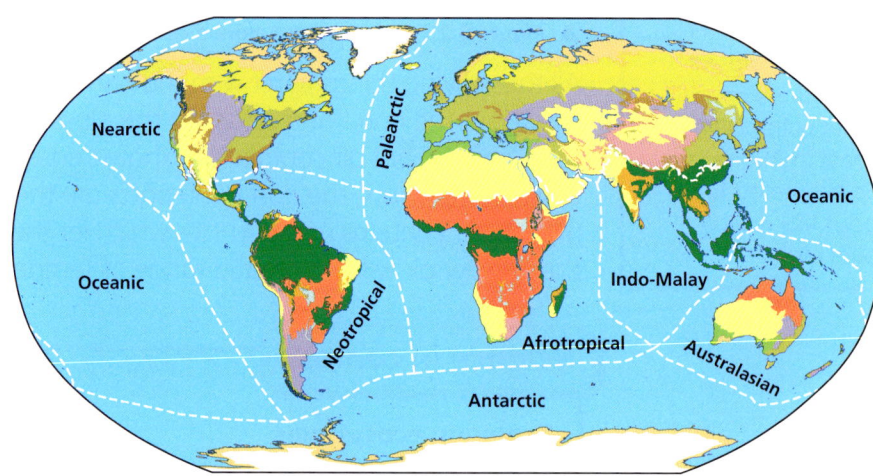

▲ **Figure 6.5** The biogeographical realms of the Earth: each realm is divided up into a large number of huge ecosystems. Each one has a distinct community of plants and animals and a certain type of climate.
Source: Millennium Ecosystem Assessment

While the idea of realms continued to develop, a German zoologist called Karl Möbius (1825–1908) investigated oyster banks (places that are set up to breed and grow oysters for food) on the German coastline to see how they may be farmed. From the results of his investigations he showed how living things interact with each other in their communities.

The idea of interaction gathered pace, and Ernst Haeckel (1834–1919), a German biologist, devised the word 'ecology' in the 1860s to

▲ **Figure 6.6** Harry Leung.

LET'S TALK

If you were an ecologist, where would you like to explore in the world and why?

describe the relationships between animals, other living things in the environment and the features of the environment such as the weather and types of soil. In the following decades, interest in such relationships gradually grew and by the beginning of the twentieth century, **ecology** – the study of living things in their environment – was established as a science, and the detailed study of ecosystems began.

Harry Leung continues the tradition of exploring, and surveying continues today with ecologists exploring all over the world! Harry Leung is a researcher of amphibians which are in danger of becoming extinct in the ecosystems of south-east Asia.

Conservation ecologist Hong Liu studies plants around the world to identify those that are in danger of extinction, and looks at the ways that they can be conserved in ecosystems in India, China, Mexico and in North America.

▲ **Figure 6.7** Hong Liu (left), with two of her students.

A vocabulary of ecology

There are a number of special terms that ecologists use frequently. You will have discussed some of them in the *Do you remember?* section of this chapter, as you may have been using them for a while. Here are the terms again, together with some others that you might find useful as you explore ecosystems and describe the living things in them.

- **Habitat –** a place where a plant, animal or microbe lives.
- **Community –** all the living organisms that live in a habitat.
- **Environment –** the surroundings of a living thing. A complete description of an environment includes all the living things, rocks, soil and weather conditions.
- **Ecosystem –** a community of living things and their environment, and the way they are linked together through the movement of nutrients and energy (as shown in food chains and webs). An ecosystem can be divided into many habitats.
- **Biodiversity –** a term used to describe the number and variety of species in an ecosystem.
- **Herbivore –** an animal that only eats plants.
- **Carnivore –** an animal that only eats other animals.
- **Omnivore –** an animal that eats both plants and animals.
- **Predator –** an animal that feeds or preys on another animal.
- **Prey –** an animal that is eaten by or falls prey to a predator.
- **Producer –** an organism that produces food at the beginning of a food chain (usually a plant).

- **Consumer –** an animal that eats plants, other animals or both.
- **Primary consumer –** an animal that eats plants (this may be a herbivore or an omnivore).
- **Secondary consumer –** an animal that eats primary consumers (this may be a carnivore or an omnivore).
- **Tertiary consumer –** an animal that eats secondary consumers (this may be a carnivore or an omnivore).
- **Quaternary consumer –** an animal that eats tertiary consumers.
- **Top carnivore –** the animal at the end of the food chain.
- **Food chain –** a description (often a diagram) of the way some organisms in a habitat are linked to each other through feeding. A food chain begins with a producer and is followed by one or more consumers, for example:

  grass → gazelle → lion

- **Food web –** a description (often a diagram) of how a number of food chains in a habitat are linked together to show how food and energy pass through the habitat.
- **Keystone species –** a plant or animal species which helps the survival of a large number of other species in a habitat. It may be a tree which provides food and resting places for animals and support for other plants, or a predator that keeps down the numbers of prey which, if allowed to increase, would reduce the number of plants in a habitat.

Three examples of ecosystems

The rainforest ecosystem

Rainforests are found on land around the middle of the planet, as Figure 6.8 shows. They are found in parts of Central and South America, Africa, Australia and Asia.

6 Which area of rainforest is closest to where you live? Try to find out more about it in the *Challenge yourself* on page 60.

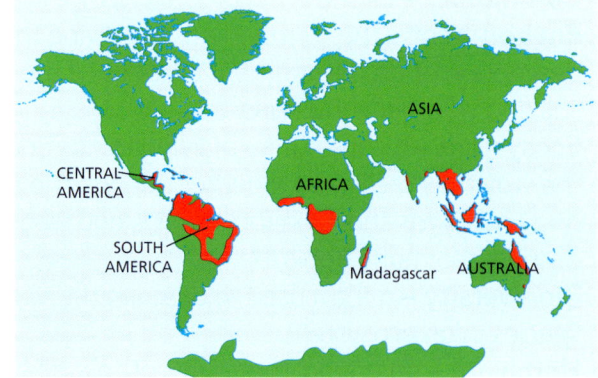

▲ **Figure 6.8** The location of rainforests around the world.

A rainforest is an ecosystem in which a community of plants, animals and microorganisms live in a hot, wet environment inside the forest. When you stand inside a rainforest, all you can see are lots of plants, as shown in Figure 6.9. Ecologists have divided the plants into four layers as Figure 6.10 shows.

▲ **Figure 6.9** A rainforest ecosystem.

▲ **Figure 6.10** The four layers of a rainforest ecosystem.

The plant layers and the soil on the forest floor form the habitats of all the living things in the rainforest.

In the **soil**, microbes decompose the dead plants and animals quickly in the warm, wet conditions. The plant roots quickly take up water and nutrients to make new growth and reproduce.

▲ **Figure 6.11** Insectivorous pitcher plants.

On the **forest floor** are many fungi, decomposing fallen trees and dead animals. Animals such as wild pigs and deer live here. There may be leeches (invertebrates that are classified into the same group as earthworms, which have suckers at each end of the body and feed on other invertebrate by sucking blood) and ticks (small invertebrates that are classified into the same group as spiders, which attach themselves to prey and feed on their blood) waiting on the leaves to climb onto any passing animal and suck their blood. There may be insectivorous plants, called pitcher plants, in this habitat.

The **understory** is formed by small trees which may be growing up into the higher layers. There may be pitcher plants here, which feed on insects growing on the trees. Birds may nest here and snakes search for prey. Larger animals like the leopard may rest here and draw up their prey from the forest floor.

▲ **Figure 6.12** Rainforests are home to a wide range of animal and insect species.

The **canopy** is formed by the branches of larger trees. They support other plants, such as orchids and ferns. This is the habitat of monkeys, such as the langur and macaque. Some rainforests have gliding frogs in their canopy, which can swoop between the branches using their webbed feet as parachutes.

The **emergent layer** is made by the tallest trees. Eagles and hawks nest in this layer and bats may roost there.

The desert ecosystem

A **desert** is an area of land which receives very little rain. There are hot and cold deserts. Deserts are found in North and South America, Africa, Australia, Asia and Antarctica.

7 Which area of desert is closest to where you live? Try to find out more about it in the *Challenge yourself* on page 61.

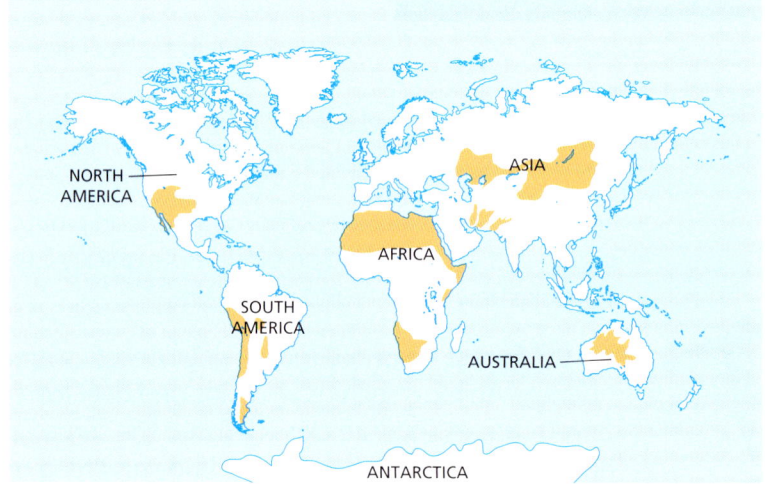

▲ **Figure 6.13** The location of deserts around the world.

▲ **Figure 6.14** Deserts are home to many species of lizard.

There are three main habitats in a hot desert. There are huge piles of moving sand called **dunes**. There are areas of rocks and thin soil, and places where water comes up through the ground to make a well or pool – called an **oasis**.

Sand dunes change positions and shapes as they are blown by the wind, and are dry for long periods. Dead insects falling on the sand form food for snakes and lizards that can burrow in the sand and then come out and move across it. If rain falls, seeds in the sand can germinate and produce plants with flowers that attract more insects.

Where there are rocks and soil, large, more permanent plants can grow, such as the cactus or the yucca. These plants provide homes for animals, such as a bird called the cactus wren, or food for mammals such as ground squirrels.

Many plants grow in the wet ground around an oasis, and are fed on by grasshoppers and butterflies, which in turn are eaten by lizards and birds.

Larger animals living in the desert, such as foxes and antelopes, visit oases for water and food.

CHALLENGE YOURSELF

Use the internet to find out about the plants and animals that live in the desert. You may like to choose the desert nearest to where you live. Make a presentation of what you find and keep it for later, when you study the effects of humans on the planet next year.

▲ **Figure 6.15** Oases provide an important water source for larger animals.

The ecosystem of the oceans and seas

The oceans and seas of the world are shown in Figure 6.16.

▲ **Figure 6.16** The oceans and seas of the world.

8 What is the nearest sea or ocean to where you live? Try to find out more about it in the *Challenge yourself* on page 63.

The ecosystem of the ocean is divided up into zones. The major ones are shown in Figure 6.17 on the next page. There may be many habitats in each zone.

The **intertidal zone** is where the tide moves up and down the sea-shore. There are two main kinds of habitat here: the rocky shore and the sandy shore. The plants and animals that live here are adapted to living out of sea-water for a few hours every day.

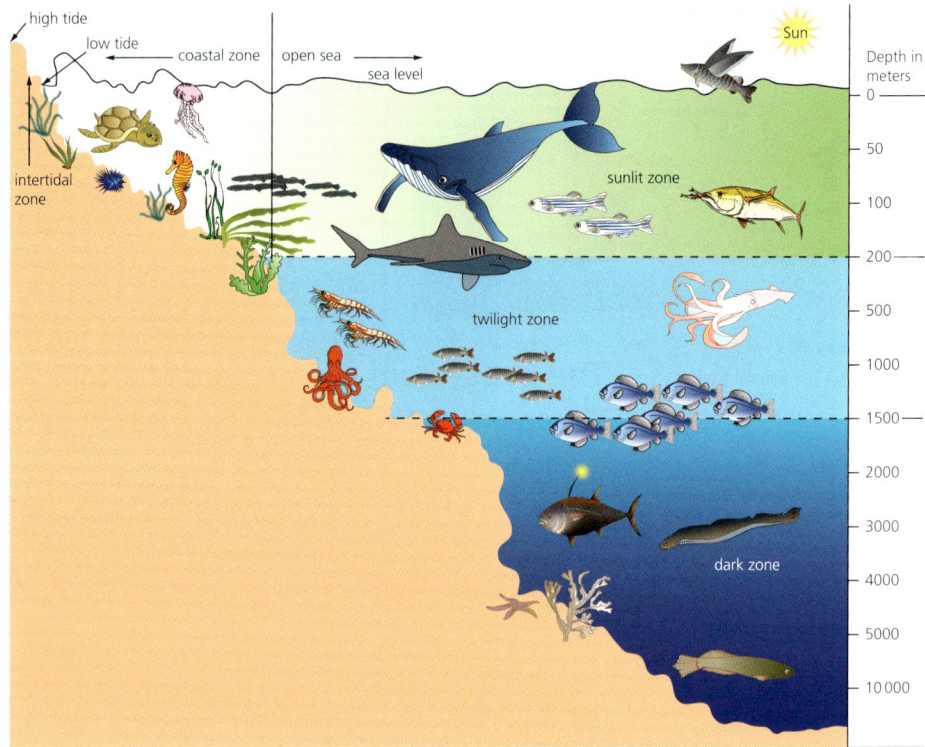

▲ **Figure 6.17** Zones of the ocean.

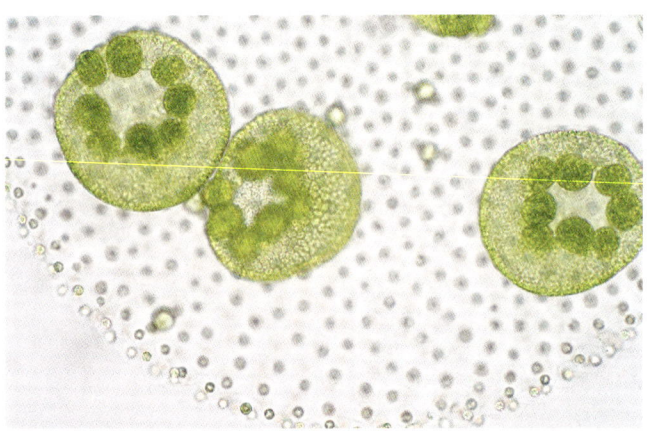

▲ **Figure 6.18** Phytoplankton.

▲ **Figure 6.19** A cuttlefish with light-emitting organs.

The **coastal zone** is further away from the shore. It has deeper water and there are no tides here. A wide range of seaweeds and animals, such as sea urchins and jellyfish, live here.

The **open sea** is far away from the coast and is divided into three zones. In the **sunlit zone**, there is enough light entering the water for microscopic plants living there to make food. These plants form part of the plankton which also includes tiny animals that feed on them.

In the deeper **twilight zone**, there is not enough light for any plants to make food. Some animals that live here feed on the dead bodies of other animals, which sink down from the sunlit zone. Other animals here are carnivores, and may have organs that light up to help them see, communicate with other animals and find food.

There is no light in the **dark zone**. The animals here feed on pieces of dead bodies that sink down from higher in the water. Those pieces that remain uneaten are broken down by decomposers on the ocean floor. There are decomposers on all the surfaces in these zones, from the sea-shore to the ocean floor, which release nutrients back into the water as they feed.

Modelling an ecosystem

From the study of many ecosystems, a simple model has been constructed that can help you understand the interactions that take place. The ecosystem can be quite small, such as a pond, or as large as a lake or a forest.

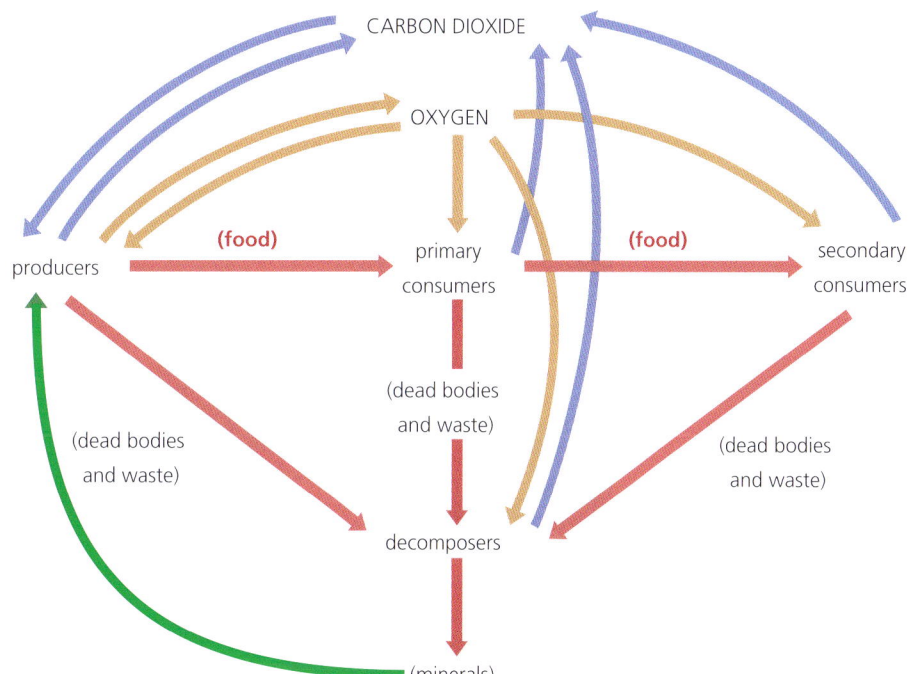

▲ **Figure 6.20** Some relationships in a simple ecosystem.

You will have already studied all of the relationships shown in Figure 6.20 in your science course, except for the arrow from producers to oxygen and the arrow from carbon dioxide to producers. You will investigate this relationship next year.

Bioaccumulation

Living things take in a variety of substances to stay alive, such as oxygen to release energy from stored food for life processes, and nutrients such as proteins to build and repair bodies. Living things can also take in **toxic substances** (also known as **toxins**) which are poisonous. These do not take part in chemical reactions to keep the body alive or to build and repair it.

9 What colour of arrow shows the movement of nutrients through food chains?

10 What colour of arrow shows the movement of minerals from the soil to the plants?

11 What colour of arrow is used to show the movement of a gas in the air that is needed to keep living things alive?

12 What colour of arrow shows the path of a gas that is produced when living things respire?

13 Where do decomposers get their food from?

Some toxins can be broken down by chemical reactions in the body and made into harmless substances which are released from the body. Other toxins may be released from the body with solid wastes in the process of egestion. If a toxin is not broken down or egested it becomes concentrated in the body by a process of **bioaccumulation**. When this happens, the toxins can damage the life processees taking place in cells, and as they build up further, the toxins eventually stop the life processes altogether and the plant or animal dies.

Some of the most common toxins are those used in **pesticides**. A pesticide is a substance, a spray or powder, that is used to kill organisms which compete with our crops and reduce the production of food. Examples of pesticides include **herbicides** (which kill weeds), **fungicides** (which kill fungi) and **insecticides** (which kill insects). If these are not used carefully, they spill over from the crop fields into the surrounding habitats and are taken up by the plants and animals living there.

▲ **Figure 6.21** Dead insects at the side of a field – killed by pesticides.

CHALLENGE YOURSELF

Is pollution damaging the ecosystems in your country? Search the internet to find out. Look at several websites. Do they provide similar or conflicting information? Can you identify any report that is biased in some way? If you can, explain your reasons. Make a presentation of your findings.

Science extra

Animals with a non-lethal level of toxins in their body may still move around in their habitat and be eaten by a predator. As the predator moves through the habitat, feeding on the toxic animals, its own level of body toxins rises higher than the levels of its prey. This increase in toxins as you go up the food chain is called **biomagnification**.

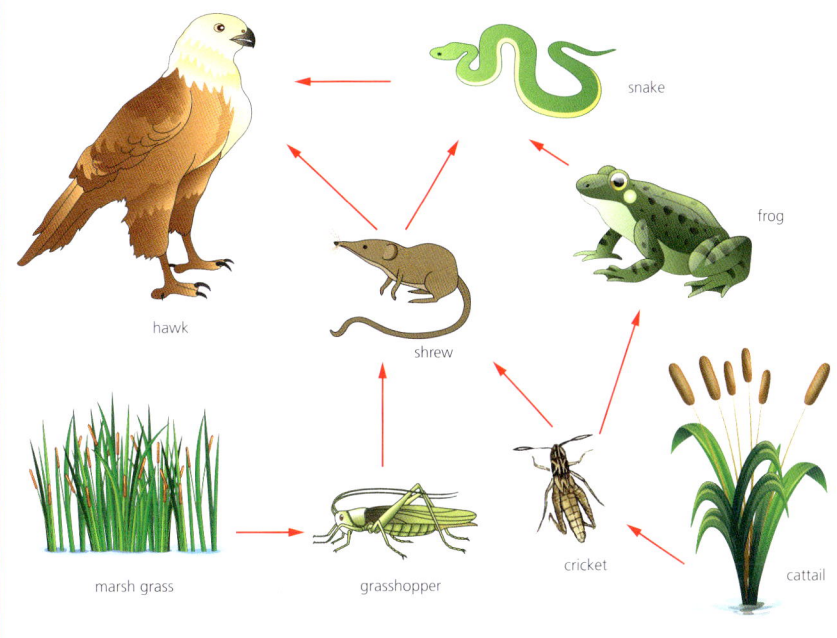

▲ **Figure 6.22** Biomagnification in a food chain.

Science in context

A poison in the food chain

The discovery of bioaccumulation and biomagnification of toxins in the food chain was made by careful ecological investigations carried out in the mid-twentieth century.

In 1935, a Swiss chemist called Paul Müller (1899–1965) set up a **research programme** to find a substance that would kill insects but would not harm other animals. Insects were his target because some species were plant pests and devastated farm crops, and others carried microorganisms that caused disease in humans. The substance also had to be cheap to make and not have an unpleasant smell. In 1939, he tried a chemical called dichlorodiphenyltrichloroethane (DDT), which was first made in 1873. DDT seemed to meet all the requirements, and soon it was being used worldwide.

In time, some animals at the end of the food chains (the top carnivores) in the habitats where DDT had been sprayed to kill insects were found dead. The concentration of the DDT applied to the insects was much too weak to kill the top carnivores directly, so investigations into the food chains had to be made.

In Clear Lake, California, DDT had been sprayed onto the water to kill gnat larvae (a gnat is a very small fly which lays its eggs in water and the larvae develop there before pupating into adults). The concentration of DDT in the water was only 0.015 parts per million (ppm), but the concentration in the dead bodies of fish-eating water birds, called western grebes, was 1600 ppm. When planktonic organisms in the water were examined, their bodies contained 5 ppm and the small fish that fed on them contained 10 ppm.

◀ **Figure 6.23** Western grebes suffered at Clear Lake in California, USA, because DDT sprayed to kill insect larvae accumulated in their bodies at high concentrations.

It was discovered that DDT did not break down in the environment but was taken into living tissue and stayed there. As the plankton in the lake were eaten by the fish, the DDT was taken into the fishes' bodies, and it built up after every meal. The small fish were eaten by larger fish, in which the DDT formed higher concentrations still. The grebes ate the large fish and with every meal increased the amount of DDT in their bodies until it killed them.

In the UK, the peregrine falcon is a top carnivore in a food chain in moorland habitats, although it visits other habitats outside the breeding season. The concentration of DDT in the bodies of female falcons caused them to lay eggs with weak shells. When parents incubated the eggs, their weight broke the shells and the embryos died.

Pollution in an ecosystem

Pollution can lead to bioaccumultation.

The most harmful pollutants in water are PCBs (polychlorinated biphenyls) and heavy metals such as cadmium, chromium, nickel and lead. In large concentrations these metals can damage many of the organs of the body, and can cause cancers to develop. PCBs are used in making plastics and, along with mercury **compounds**, are taken in by living organisms at the beginning of food chains. They are passed up the food chain as each organism is eaten by the next one in the chain. This leads to organisms at the end of the food chain having large amounts of toxic chemicals in their bodies, which can cause permanent damage or death.

14 Construct the food chain investigated in Clear Lake.

15 Why did the grebes die?

16 How are the lives of people who live by polluted rivers, and catch fish from them, put at risk?

17 The water flowing through a village had such low levels of mercury in it that it was considered safe to drink. Many of the villagers showed signs of mercury poisoning. How could this be?

Pollution can alter the relationship between organisms in an ecosystem.

Fertilizers are used to make crop plants grow faster and larger. The careless use of fertilizers allows them to drain from the land into rivers and lakes, and leads to the overgrowth of water plants, including algae. This is called algal bloom. When these plants die, large numbers of decomposing bacteria take in oxygen from the water, and this reduction in oxygen levels in the water kills many water animals. Phosphates in detergents also cause an overgrowth in water plants, which can lead to the death of water animals in the same way.

▲ **Figure 6.24** The excessive use of fertilizers leads to algal blooms in rivers and kills fish.

Pollution can physically damage organisms in an ecosystem.

Litter, in the form of tins and plastic, can cover up plants and prevent light getting to them, so they die.

◀ **Figure 6.25** Litter can cause lasting damage to ecosystems.

Glass bottles can become traps for small animals like mice. Once they have crawled inside, they cannot grip the smooth walls inside the bottle and climb out. Plastic bags can be eaten by a wide range of animals. The plastic can't be broken down and blocks the digestive system, stopping it from working, and the animal dies.

▲ **Figure 6.26** Plastic can exist in ecosystems for a long time and cause huge damage.

Invasive species

An **invasive species** is a species of living thing which enters an ecosystem in which it is not naturally found and causes damage to that ecosystem.

An early example of the effect of invasive species occurred during the European colonisation of Australia, when European rabbits were introduced to ecosystems which had developed naturally over long periods of time on the Australian continent. Rabbits breed quickly, but in Europe their numbers are controlled by predatory animals like foxes. Without foxes to control numbers, the rabbit population rapidly increased, so foxes were then introduced to prey on them. **Marsupial** mammals live in the ecosystems of Australia and have a slower behaviour than rabbits, so the foxes found them easier prey. Domestic cats brought over by the settlers also found some native mammals easy prey, and the invasion of foxes and cats led to the extinction of over twenty Australian mammal species. The populations of some mammals, such as the boodie in Australia, are threatened by invasive species.

▲ **Figure 6.27** A boodie or burrowing bettong.

Pampas grass is an example of an invasive plant. It was introduced from its natural habitat in the Andes in South America into gardens in New Zealand. As its seeds can travel up to 25 km on the wind, it has spread into surrounding ecosystems where it competes for light and nutrients with the plants that grow there naturally.

▲ **Figure 6.28** Pampas grass growing in a wild habitat in New Zealand.

Summary

✔ An ecosystem is the interaction of a collection of animals, plants and microorganisms with each other and in a specific place.

✔ Science in context: There is a long history of scientists, from Alexander Humboldt 250 years ago to Harry Leung in the present day, exploring ecosystems around the world.

✔ There are many different ecosystems on the Earth, including rainforests, deserts, and oceans and seas.

✔ We can understand ecosystems visually by modelling them.

✔ Bioaccumulation is the process of the build-up of toxic substances on an ecosystem and can have a harmful impact on ecosystems and living things.

✔ Science in context: Ecological investigation led to a greater understanding of the damage certain substances, such as DDT, caused to ecosystems and food chains.

✔ Invasive species are those that are new to an ecosystem, many of which can affect other organisms and the ecosystem.

End of chapter questions

1 Name two ecosystems found on land on the Earth

2 Name four habitats in the ocean ecosystem.

3 What is a toxin?

4 How are toxins harmful to living things in an ecosystem?

5 Name two toxins which can damage an ecosystem.

6 What does 'bioaccumulation' mean?

7 What is an invasive species?

8 Give an example of an invasive species.

9 How can an invasive species damage an ecosystem? You may use one or more examples in your answer.

 Now you have completed Chapter 6, you may like to try the Chapter 6 online knowledge test if you are using the Boost eBook.

Investigating an ecosystem

In this chapter you will learn:
- about habitat surveys (Science in context)
- how to plan a range of investigations into habitats around you (habitat surveys)
- how to consider variables for your investigations and collect and record sufficient observations
- how to make predictions of likely outcomes based on your understanding of habitats
- how to present the findings of your investigations.

Do you remember?

- What is an ecosystem?
- What is a habitat?
- What can damage an ecosystem?
- How can this damage affect living things in the ecosystem?

Investigating an ecosystem

Ecologists make surveys of a habitat to find out about the ecosystem – the way the plants and animals interact together. They do this by making surveys to identify the species present and also to estimate the number of individuals in each species. During this survey they may closely observe the plants and animals and note how they are adapted to their habitat or how they interact with each other.

Later, a second survey is made and the data collected is compared with the data from the first survey. This allows the scientists to see if there has been a change in the number of species or a change in the estimated number of each species. These changes give an indication of how the plants and animals are interacting. This leads to assessing the state of the community – whether it is stable (remaining in the same condition) or becoming unstable (changes taking place in numbers of species and individuals in a species). If there are signs of instability, further investigations need to be made to find the reason for this change. The data for these surveys is collected by a range of surveying techniques.

1 When the earliest explorers surveyed a habitat for plants, how do you think they might have categorised the species they found based on the evidence they observed?

2 When the earliest explorers surveyed a habitat for animals and saw them moving around, how do you think they might have categorised the different species based on the evidence they observed?

3 How reliable was the evidence from the first expeditions? Explain your answer.

4 How reliable was the evidence from the later expeditions? Explain your answer.

LET'S TALK

What television programmes about wildlife have you seen recently? How good are they at providing information about habitats and the lives of plants and animals? Explain your answer. In what ways might the programmes make people concerned about what they see?

Science in context

Habitat surveys

People have been surveying habitats since the earliest times. The purpose was to find fruit, seeds and roots, as well as animals, such as mammals and birds, to eat. Much later, when countries in Europe sent out sailing ships to explore the world, habitat surveys were made at each new land that was visited. They were not called habitat surveys. They were just journeys through the new land where the first explorers, often with no scientific knowledge, observed and collected specimens. The explorers wrote down their observations and brought specimens home, by which time they were usually dead and in some form of decay. However, the arrival of specimens from new lands fired the interest of scientists, and eventually the expeditions became more scientific. Artists took part to draw specimens in their habitats, and scientists recorded observations more systematically in notebooks and preserved dead specimens in bottles of alcohol for the journey home.

In the 1950s, television companies sent out expeditions in which the plants and animals of a habitat were recorded on film. Wildlife films like this are still being made today and stimulate many people to take up a career in biology.

There is a second type of habitat survey taking place today which is much more scientific. The aim of these surveys is to describe the habitat as accurately as possible for two reasons. First, the habitat might be that of a rare or endangered species, so the purpose is to discover how the activities of the species link with other species and thus work out the best way of setting up a conservation programme. Second, the habitat may be one that is under threat from the human population; the land on which it stands may be being considered for building houses and factories, or for flooding to make it into a reservoir for a hydro-electric power station, for example

So how are habitat surveys made today? Here are the stages that you would go through if you were a scientist carrying out such a survey.

The chances are that even the most remote habitat has been visited by scientists in the past and there may be records of their observations that can be examined. This process of looking at recorded data and accounts of previous journeys in the habitat is called the ecological desk survey.

Next, the scientists visit the habitat and make a survey in which they note down its features and make a map. They may also talk to people who live locally about the plants and animals they have seen. Often these local people have a different language from the scientists and may have several names for a plant or an animal.

After the survey, transects are set up over the habitat. Transects are lines made across an area in an ecological survey, often by using a rope with certain intervals marked along it, at which observations and recordings are made. Along each transect, at regular intervals, stations are set up at which the scientists record what they can see and hear. They assess the numbers of each species at the station using the DAFOR code – the letters stand for Dominant, Abundant, Frequent, Occasional, Rare. On the walk out along the transect, the abundance of the different species of animals might be recorded along with their activities. On the return walk, the abundance of plants might be noted together with their condition – for example, the plants might be 'in flower' or 'bearing fruits'. If there is a team of scientists, each one may walk a transect at different times during the day.

▲ **Figure 7.1** These scientists are using an insect trap as part of a habitat survey in French Guiana.

5 Could a desk survey provide inaccurate evidence? Explain your answer.

6 How accurate do you think evidence given by local people might be? Explain your answer.

7 How do you think the evidence provided by using a transect for one day and one night might be made more reliable?

8 Make a plan for a local habitat survey, perhaps part of your school grounds, to describe the habitat as accurately as possible.

Night surveys are also taken – this is when the scientists walk the transects in darkness and note the animals heard or seen by flashlight. A bat recorder may be switched on to record the sounds of bats, which humans cannot hear. If insect traps have been set up at the stations, these can be examined. At some stations, photo traps can be set up, in which a camera triggered by a harmless trip-wire records an image of any animal that passes by after the scientists have gone.

Surveys like these are taking place all over the world right now. One example is along the Kenyan coast where rainforest surveys are being made in the study of black and white colobus monkeys, which are an endangered species.

▲ **Figure 7.2** Information about the habitat of black and white colobus monkeys in the Kakamega Forest Reserve in Western Kenya could help to protect this **vulnerable** species.

Making a survey

LET'S TALK

Select an ecosystem to investigate. It may be an area in your school grounds or a local park. Share your ideas with the rest of the class and make a joint decision on the ecosystem you will investigate. See if there is any previous data collected by other teams in the past that will allow you to make a desk survey (see below).

The stages in a survey

1 The ecological desk survey

If the ecosystem has been surveyed before, there should be some data collected by the previous ecologists. This data could be examined to see the different types of living things (or species) that were present then (see *Can you learn more from previous data?* on the next page). From this, you may modify the surveys suggested later in the chapter to fit in with the sampling from the past.

Can you learn more from previous data?

If previous data is available, there may be a map of the habitat and an assessment of the plants and animals there using the DAFOR code (see page 73). You may use this information when planning the survey for the teams in the class.

2 The map

Here are some surveys you can make in a habitat to find out about its ecosystem. You do not have to try them all, but you do need to work out as a class how you will investigate the ecosystem.

Can you make use of previous data?

If there is previous data then there may also be a map. If you need to map your habitat, see if an outline is available from the school office, or look on a map of your chosen area. Alternatively, look at a map on the internet and zoom in to show the position of features, such trees and paths, and make a hard copy.

3 Making the survey

You will need to use a range of survey techniques, as shown on the following pages. Divide the class into teams so that each one can use one of the survey techniques (see the *Let's talk* on page 79).

Surveying techniques

If you are to make science enquiries into ecosystems, you need to know about the various survey techniques you can use.

Observing and recording plant life

Using a quadrat

The quadrat shown in Figure 7.3 is square, but a quadrat can be any fixed shape which encloses a known measured area. This quadrat is placed on the ground, but transparent plastic quadrats can be placed on a tree trunk to observe the lichens and moss living there.

Making a transect

If there is a feature such as a bank, a footpath or a hedge in a habitat, the way it affects plant life can be investigated using a line transect. This is made by stretching a rope along a line that you wish to examine, then observing and recording the plants growing at certain intervals, called stations, along the rope.

▲ **Figure 7.3** Using a quadrat to map the daisies in a lawn.

▲ **Figure 7.4** Making a transect down a hillside.

Collecting small animals

Collecting from soil and leaf litter

A Tullgren funnel can be made from laboratory equipment as shown in Figure 7.5. The sample of soil or leaf litter is placed on the gauze inside the paper collar and the lamp is brought close, but not touching, and then switched on. Animals in the sample will move away from the heat and fall into the beaker, where they can be examined before being returned to their habitat.

paper collar (prevents insects walking away)

leaf litter

gauze

funnel

beaker

moist tissue paper

▲ **Figure 7.5** A simple Tullgren funnel.

The pitfall trap

The pitfall trap is used to collect small animals that move over the surface of the ground. It is made by digging a hole in the ground, and placing a yoghurt pot containing few small leaves inside it. Four pebbles are arranged around the hole and a piece of wood or plastic is placed on top to keep rain out of the pot. Animals may move between the pebbles and fall in. After you have observed and recorded the animals, they should be returned to their habitat.

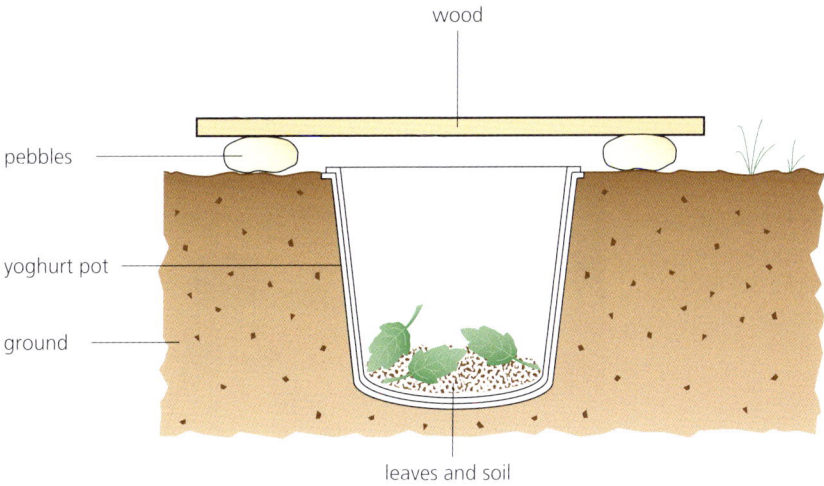

▲ **Figure 7.6** A pitfall trap.

Sweep net

The sweep net is used to collect small animals from the leaves and flower stems of herbaceous plants, especially grasses. Herbaceous plants do not have stems made of wood, so at the end of the growing season the stem dies and a new one grows in the next season. Figure 7.7a shows how to hold the sweep net with the lower edge slightly forwards. The net is then moved in a semicircle keeping the lower edge slightly forwards so that it scoops up any small animals on the plants. After one or two sweeps, the mouth of the net should be closed by hand and the contents emptied into a large plastic jar where the animals can be identified. The animals must then be returned to their habitat.

▲ **Figure 7.7** Using a sweep net.

Sheet and beater

Small animals in a bush or tree can be collected by setting a sheet below the branches and then shaking or beating the branches with a stick. The vibrations dislodge the animals, which then fall onto the sheet. Large animals such as beetles and snails can be observed and recorded on the sheet before being placed at the base of the tree or bush. The smallest animals can be collected in a pooter (see Figure 7.8).

Pooter

Tube A of the pooter is placed close to the animal and air is sucked out of tube B. This creates low air pressure in the pooter so that air rushes in through tube A, carrying the small animal with it.

9 What is the purpose of the cloth cover on the end of tube B inside the pooter?

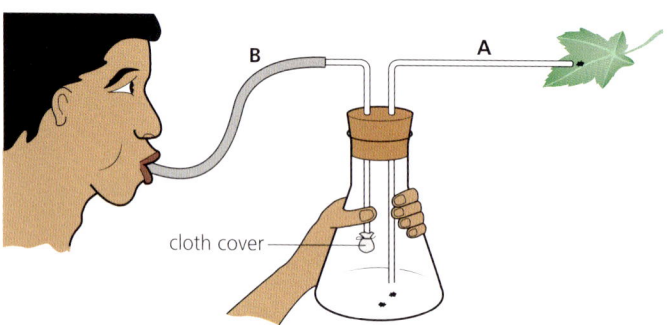

cloth cover

▲ **Figure 7.8** Using a pooter.

Collecting pond animals

Pond animals may be collected from the bottom of the pond, from the water plants around the edges or from the open water just below the surface.

A drag net is used to collect animals from the bottom of the pond. As it is dragged along, it scoops up animals living on the surface of the mud.

The pond-dipping net is used to sweep through vegetation around the edge of the pond, to collect animals living on the leaves and stems.

A plankton net is pulled through the open water to collect small animals swimming there.

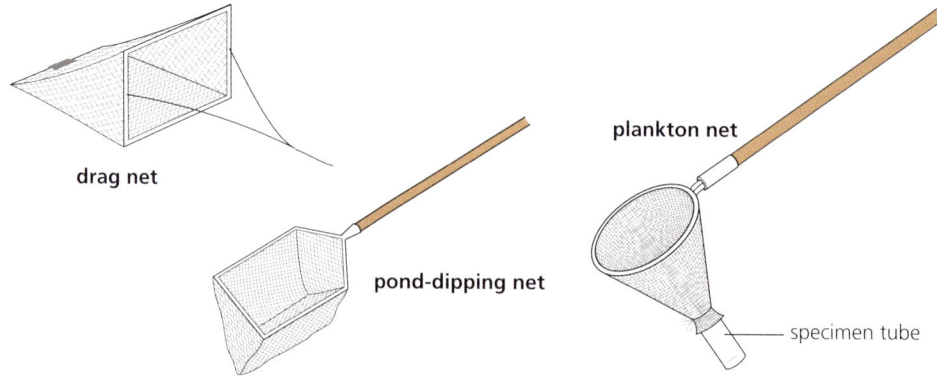

drag net

plankton net

pond-dipping net

specimen tube

▲ **Figure 7.9** Three types of pond net.

Surveys to make

Previous surveys of your chosen habitat may include investigations of the following:

- plants in an area of short grass
- plants growing under a tree
- small animals in leaf litter and soil
- small animals moving over the ground
- small animals resting on taller grass
- small animals resting on the branches of trees or bushes.

If you have previous data, you may like to add to it by repeating the survey. If there is no previous data for your chosen habitat, try the Survey enquiries 1–6 on pages 79–83.

LET'S TALK

Look through the six surveys, then hold a class discussion and set up teams who will each use one of the techniques to add to previous data, or to try one of the enquiry surveys. Work out with your team the various tasks to be carried out; for example, using the equipment, identifying the types of plants and animals (such as 'plant with yellow flowers', 'green beetle'), and recording the numbers of organisms examined. Decide if you would like to rotate these tasks during the survey, but consider the effect this might have on keeping the test fair.

Warning

Some ecosystems have poisonous plants and animals. These surveys must only be made where it is safe. Check with your teacher and have your investigating site approved before you make your surveys. Also check with your teacher that that you are wearing suitable clothing for the ecosystem, which may include hats, gloves, long sleeves and long trousers tucked into socks.

Survey 1: Do other plants grow in an area of short grass?

You will need:

a quadrat, paper, a pen and a clipboard on which to record your results.

Investigation and recording data

1 Place a quadrat over an area of short grass and look for plants that are not grass. They will have leaves which are wider, shorter or a different shape from grass leaves.
2 You can supply a simple answer by counting the different types of other plants you find. Make a record of what you find.

Examining the results

Look at the data you have collected. Do all samples show the same result?

Conclusion

Use your results to answer the survey question.

Is your conclusion limited in some way? Explain your answer.

What improvements could be made? Explain the changes that you suggest.

Survey 2: Do the branches of a tree overhanging a grassy area affect the plants that grow there?

You will need:

a long rope, a metre rule, coloured paper or ribbon to mark the stations along the rope, paper, a pen and a clipboard on which to record your results.

Investigation and recording data

1 Set up a line transect from the base of the tree trunk, out across the grassy area under the tree branches, and continue to set out the rope for three metres into the open air.
2 Set up stations every half-metre (50 cm) along the rope, and then observe and record the plants growing at each station.
3 You can supply a more detailed answer by setting up the stations 20 cm apart, and observing and recording the plants that are growing there.
4 You can check your data by setting up transects at 30° and 60° on either side of the first one you made.

Examining the results

Look at the data at each station and see if there is a trend, a pattern or an anomalous result.

Conclusion

Use your results to answer the survey question.

Is your conclusion limited in some way? Explain your answer.

What improvements could be made? Explain the changes that you suggest.

Survey 3: Does leaf litter from different trees have different animals living in it?

You will need:

gloves, the equipment for making a Tullgren funnel (to set up in the laboratory), paper and a pen.

Investigation and recording data

1   Use gloves and a bag to collect leaf litter from under one type of tree. Note the position of the tree in the ecosystem.
2   Set up a Tullgren funnel without the lamp and place your leaf litter sample inside the paper collar.
3   Carefully lower the lamp close to the sample but not touching it. Switch on the lamp and leave it for an hour.
4   Look at the paper in the beaker and, if there are some animals in it, switch off the lamp and remove it.
5   Observe and record the animals in the beaker, then return them to the leaf litter and to the ecosystem.
6   If there are no animals after an hour, leave for another hour and observe again.
7   Repeat steps 1–6 with leaf litter from another type of tree.

Examining the results

Compare the data from the two samples.

Conclusion

Use your results to answer the survey question.

Is your conclusion limited in some way? Explain your answer.

What improvements could be made? Explain the changes that you suggest.

10   Why do you think the animals move down through the leaf litter and through the holes in the gauze?

11   How could you provide more reliable data to research this question?

Survey 4: Do the same types of animals move around on soil as on a grassy surface?

You will need:

three pitfall traps.

Investigation and recording data

1   Find an area which is just soil and find an area covered by short grass.
2   Set up three pitfall traps in each area and leave them overnight.

12 Where else in the ecosystem could you set up pitfall traps to investigate further? If your teacher approves, set them up and investigate.

3 The following morning, examine the traps and observe and record any animals you find.
4 Return the animals to their habitat.

Examining the results

Compare the data from the pitfall traps in the two areas.

Conclusion

Use your results to answer the survey question.

Is your conclusion limited in some way? Explain your answer.

What improvements could be made? Explain the changes that you suggest.

Survey 5: How many different kinds of animals rest on grass about 30 cm tall?

You will need:

a sweep net and a jar.

Investigation and recording data
1 Select an area of grass and use a sweep net to make one sweep, then empty it into a jar and observe and record any animals found.
2 Repeat step 1 five times.

Examining the results

Compare the animals swept into the net at different times by answering the following questions.

Do all sweeps show the same animals?

Do some sweeps show different animals from other sweeps?

How many different animals were swept up into the net?

Conclusion

Use your results to answer the survey question.

Is your conclusion limited in some way? Explain your answer.

What improvements could be made? Explain the changes that you suggest.

CHALLENGE YOURSELF

The pooter shown in Figure 7.8 (page 78) has been constructed in a laboratory using laboratory equipment, but you can construct one from simpler pieces of equipment. Search the internet for 'how to make a pooter'. Look for the site of a wildlife organisation and discuss it with your teacher. If approved, make and test your pooter and use it in Survey 6.

Survey 6: How many different kinds of animal are resting in the branches of a bush or a tree?

You will need:

a sheet and a stick to beat the branches with.

Plan, investigation and recording data

1   Place a sheet beneath the branches on one side of a tree.
2   Beat the branches with a stick. Do not strike them too hard – just enough to make the branches vibrate a little.
3   Observe and record the animals that have fallen onto the sheet.
4   Place all the animals around the base of the trunk of the tree or bush so they may climb back into it.

Examining the results

Look at the data from the different parts of the tree and search for a trend or pattern.

Conclusion

Use your results to answer the survey question.

Is your conclusion limited in some way? Explain your answer.

What improvements could be made? Explain the changes that you suggest.

Surveys and habitat description

Make your surveys, record your data and make a presentation of what you have done. If you have made the surveys suggested here, present your data and conclusions and answer any further questions. The whole class should then take part in a scientific seminar where each survey is presented, so that everyone has information about all of the investigations and about what was discovered.

▲ **Figure 7.10** Presenting your findings is an important part of the scientific process.

CHALLENGE YOURSELF

From what you have learnt from your seminar, write a description of the ecosystem in the habitat your class has studied.

13 A second survey was made a few months after the first in a habitat where there are different seasons in the year. How reliable is the comparison of this data in a monitoring programme? Explain your answer.

The ecosystem surveys provide a snapshot of the habitat at a particular time. This information has even more value when another similar investigation is made a year later and the two results are compared. Making two surveys is the start of a longer research programme called **monitoring**. In a monitoring programme, the data collected by the yearly surveys helps to assess the stability of the ecosystem.

Taking part in an ecological monitoring survey helps people to work together to understand an ecosystem, and helps to see how it changes over time. It also helps people to see how the results obtained by different members of a group can be put together to give a better understanding of our scientific knowledge.

CHALLENGE YOURSELF

Develop your survey into a monitoring programme. Arrange to meet up with the other members of your group in a year's time and investigate again. Present your data and compare it with the data you presented in the first investigation. Look for changes in stability of the ecosystem and, if any are found, investigate further. Encourage others from lower Stages in your school to make this enquiry next year and add their data to yours. In time, over five or more years, your school could have an ecosystem monitoring programme which could be valuable to professional ecologists.

Summary

- ✔ We can plan and carry out investigations into habitats around us, called habitat surveys.
- ✔ Science in context: There are a variety of surveys ecologists can carry out, such as habitat surveys, ecological desk surveys, night surveys as well as the use of transects and other tools.
- ✔ It is important to understand and identify the variables for your investigations and to collect and record sufficient observations.
- ✔ Survey tools and methods include quadrats, transects, Tullgren funnels, pitfall traps, sweep nets, sheets and beaters, and pooters.
- ✔ We can make predictions of likely outcomes based on our understanding of habitats and present the findings of our investigations in a number of different ways.

End of chapter questions

You can test your skills in thinking and working scientifically in ecosystems by answering these questions.

1 What is a quadrat?
2 What is a quadrat used for?
3 Describe or draw a Tullgren funnel.
4 What is a pitfall trap?
5 How would you set up a pitfall trap?
6 What would you predict to catch in a pitfall trap?
7 What must you do to control risks when making a survey in an ecosystem?
8 How do you make a transect from under a tree to open grass three metres away?
9 One end of a transect was set up in a grassy area 5 m from a tree trunk. The other end was set up 5 m from the trunk on the other side of the tree. The tree's branches stretched to 2 m on either side of the trunk. A pitfall trap was set up at one end of the transect and at 1-m intervals to the other end. After 1 day, the pitfall traps were emptied. The table shows the results.

▼ **Table 7.1**

Species found	Pitfall trap										
	0	1	2	3	4	5	6	7	8	9	10
herbivorous beetles	6	6	5	2	0	0	0	1	5	6	6
carnivorous beetles	0	1	2	3	5	6	4	2	2	1	0
moth caterpillars	0	0	0	6	10	12	8	4	0	0	1
spiders	0	0	0	1	3	4	2	1	0	0	0

  a Make a bar chart for each of the animals in the table.
  b Suggest a reason for the numbers of herbivorous beetles at stations 0–5.
  c Suggest reasons for the numbers of moth caterpillars at stations 0–5.
  d Suggest reasons for the numbers of carnivorous beetles and spiders at stations 0–5.
  e How do the numbers of animals at traps 6–10 compare with those at traps 0–5?
  f Identify a result that does not fit in with the other results and offer an explanation for it.

Now you have completed Chapter 7, you may like to try the Chapter 7 online knowledge test if you are using the Boost eBook.

8 The structure of atoms

In this chapter you will learn:

- about the first ideas about atoms (Science in context)
- about investigating chemical reactions (Science in context)
- about Dalton's atomic theory (Science in context)
- about the plum pudding model (Science in context)
- about the Rutherford model of the structure of the atom
- that electrons, protons and neutrons have different charges and what these are
- that the chemical properties of elements are linked to their atomic structure
- about electrostatic force and that it is this force that holds individual atoms together.

Do you remember?

- What is an atom?
- What is an element?
- What is the periodic table?

If you have answered the questions above correctly, then you already know something about atoms, but how did our knowledge about them build up to what we know today?

Science in context

The first ideas about atoms

Democritus (about 460−370 BCE) thought about the structure of materials and what would happen if you cut something into smaller and smaller pieces until you could not cut it any smaller. He concluded that you would come to an indivisible particle, which he called an 'atom'. Building on this idea that everything was not made of water or air, but of atoms, he went on to say that atoms of different substances had different sizes and shapes, and that they were able to join together to make even more different substances.

The other **philosophers** could not agree on Democritus' creative thought and later Aristotle put forward the idea that matter was made from water, air, fire and earth. This idea was believed to be true until scientists began investigating chemical reactions.

Modelling Democritus' first atoms

You will need:

a sheet of paper and a camera (optional).

Process

Take the sheet of paper and tear it in half. Take one half of the sheet of paper and tear it in half. Take a quarter of the sheet of paper and tear it in half. Take an eighth of a sheet of paper and tear it in half, and so on until you have a piece of paper so small that you cannot tear it in half. This is your model atom.

You may like to communicate Democritus' idea by repeating the activity, filming it and providing a commentary to explain his thinking.

Considering strengths and weaknesses

How accurate is the model?

Examine the model for strengths and weaknesses by answering the following questions:

1 Does the model support the idea that atoms are small?
2 Does the model show that an atom is related to a piece of material?
3 Paper is made of the atoms of many elements. Is it a good material to use for this model? Explain your answer.
4 If a material is made of atoms of many elements, it will have properties that are different for each of the elements. If one atom was taken from the paper, would it have the same properties, like the tiny piece of paper you made? Explain your answer.

Analogy in models

What was the analogy of an atom in this model?

1 What material was used to make the analogy in this activity?

Science in context

Investigating chemical reactions

Antoine Lavoisier (1743–1794), a French chemist, investigated the changes that took place when two chemicals reacted and formed a new compound. He weighed the chemicals before the reaction and then weighed the compound that was formed. Lavoisier found that the total mass of the chemicals was the same as the mass of the compound that was produced. From this result and from the results of similar experiments, Lavoisier set out his law of conservation of mass, which stated that matter is neither created nor destroyed during a chemical reaction.

2 What piece of equipment must Lavoisier have used to discover that in a chemical reaction mass is conserved?

3 What scientific models did Marie-Anne make?

4 How did Marie-Anne bring more knowledge of science to Lavoisier's laboratory?

5 How did Lavoisier's work influence Proust's work?

6 How did Proust's work build on the work of Lavoisier?

Lavoisier was assisted in his work by his wife Marie-Anne, who made sketches of the new pieces of equipment that were devised for the investigations and translated the work of other scientists into French for Antoine to read.

Joseph Proust (1754–1826), another French chemist, followed Lavoisier's example by carefully weighing the chemicals in his experiments. He discovered that when he broke up copper carbonate into its elements – copper, carbon and oxygen – and then weighed them, they always combined in the same proportions of five parts copper, four parts oxygen and one part carbon. He found that other substances were made from different proportions of elements and these proportions were always the same too, no matter how large or how small the numbers of elements that were used. From his work, Proust devised the law of definite proportions, which stated that the elements in a compound are always present in a certain definite proportion, no matter how the compound is made.

We can see that Lavoisier and Proust were working in the same country, at about the same time, but a little later, another scientist in another country built up a theory based on their work.

Science in context

Dalton's atomic theory

John Dalton (1766–1844) was an English chemist who studied gases, and from his investigations on the combining of carbon and oxygen he produced two gases. The first of them seemed to be made from one particle of carbon joining with one particle of oxygen, and in the second gas it seemed that one particle of carbon joined with two particles of oxygen. From his own observations, and from reading about the work of Lavoisier and Proust, Dalton put together his atomic theory. He suggested that:

- All matter is composed of tiny particles called atoms.
- Atoms cannot be divided up into smaller particles, and they cannot be destroyed.
- Atoms of an element all have the same mass and properties.
- The atoms of different elements have different masses and different properties.
- Atoms combine in simple whole numbers when they form compounds.

7 Dalton used evidence from the work of others to help him develop his theory.

   a Who provided evidence for the first statement of his theory?

   b Who provided evidence for the second statement of his theory?

   c What two pieces of evidence helped him to develop the fifth statement of his theory?

▲ Figure 8.1 John Dalton.

Dalton's theory helped chemists at the time, but the results of later investigations showed that it was not completely correct, as we shall see.

The plum pudding atom

The scientists in these early chemical studies on atoms used a unit called the 'atomic weight' to compare the elements. Today, we use the term 'relative atomic mass' or RAM. An English chemist called William Prout (1785–1850) studied the atomic weights of the different elements and thought he could use them to explain the structure of atoms. He knew that hydrogen had the lowest atomic weight, and that the atomic weights of all the other elements appeared to be multiples of the atomic weight of hydrogen. This suggested to him that all the other elements were made up from different numbers of hydrogen atoms. This idea was later shown to be completely wrong, but it did make scientists such as Joseph J Thomson (1856–1940) think that atoms might have a structure inside them.

During the nineteenth century, great developments were made in the study of electricity and the development of electrical equipment for use in investigations. One of these pieces of equipment was the **cathode ray tube**, which produces rays when it is connected into an electrical circuit. The ray is produced from the material from which the cathode is made. Thomson investigated these rays and discovered that they were made of tiny particles, which had a mass over a thousand times smaller than a hydrogen atom. Thomson called the particles 'corpuscles', but George Stoney (1826–1911), an Irish physicist, named them **electrons**. When he used different materials for the cathode, he always found that the electrons they produced were the same.

In 1904, Thomson devised a model of the structure of the atom from his studies on electrons. He proposed that electrons were present in the atom. He knew that electrons were negatively charged and atoms were neutral, so the negatively charged electrons must be balanced by a positive substance in the atom. He described the atom as being like a plum pudding, with the negatively charged electrons being surrounded by a positively charged 'pudding'.

8 Why could Prout's idea be considered a creative thought?

9 Did the evidence produced by Thomson's investigations with the cathode ray tube support Dalton's atomic theory? Explain your answer.

10 What creative thought did Thomson have after his discovery of electrons?

▲ **Figure 8.2** Joseph J Thomson at work in his laboratory.

11 In the plum pudding analogy, what do the currants and raisins represent?

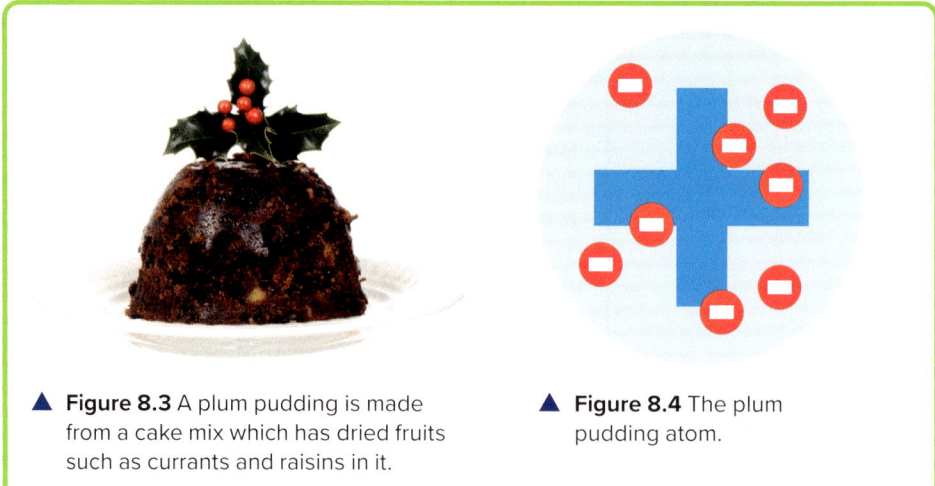

▲ **Figure 8.3** A plum pudding is made from a cake mix which has dried fruits such as currants and raisins in it.

▲ **Figure 8.4** The plum pudding atom.

CHALLENGE YOURSELF

In the plum pudding analogy, the cake mix and dried fruits were used as analogies for part of the atom. Think of other objects that could be used as analogies for part of the atom. Discuss them with your teacher and, if approved, make a three-dimensional model of your analogy of the structure of the atom.

CHALLENGE YOURSELF

Look at the diagram of the plum pudding atom shown in Figure 8.4. How could you turn it into a 3-D model? Write out your plan and, if your teacher approves, make it.

What materials did you use to construct your analogy?

Ernest Rutherford and the atom

Ernest Rutherford (1871–1937) was born and raised in New Zealand and, after his successful studies on electricity and magnetism at the University of New Zealand in Wellington, he moved to Cambridge University in England to work with JJ Thomson. He spent time studying radioactive materials and the radiation that they produced with Paul Viliard (1860–1934), a French scientist. Between them, they discovered that there are three types of radiation – **alpha particles**, **beta particles** and **gamma rays**.

12 What scientific enquiry skill was Rutherford using in selecting Thomson's model for investigation?

13 What equipment did Rutherford select for his investigation?

Rutherford had found that alpha particles were large, positively charged particles, much bigger than electrons, and he decided to use them to test Thomson's idea about the plum pudding structure of the atom.

Rutherford's plan was to hang up a thin sheet of gold and surround it with a screen that could detect alpha particles, as shown in Figure 8.5. He would then fire alpha particles at the sheet and the alpha particles would

14 What observations were necessary in the investigation?

15 What did Rutherford predict if Thomson's model was correct?

eventually hit the screen and be detected. From the marks made by the alpha particles on the screen, he could work out the structure of the atom. Rutherford predicted that if Thomson's model was correct, all the alpha particles would pass straight through the gold atoms and make a mark directly behind the gold sheet.

The experiment

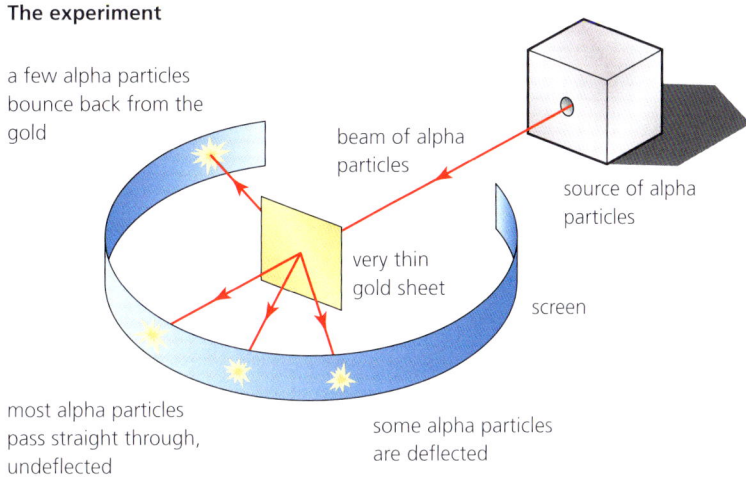

Explanation

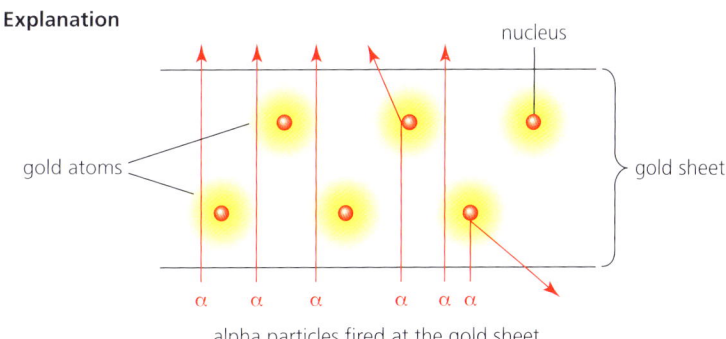

alpha particles fired at the gold sheet

▲ **Figure 8.5** Rutherford's experiment investigating atomic structure.

The experiment was set up and the alpha particles were fired at the gold sheet. Most of the alpha particles made a mark directly behind the metal sheet, but some made marks all round the screen. This did not fit in with the prediction, and suggested that the atoms had a structure that was not like a plum pudding.

16 How did the results compare with the prediction?

17 What creative thought did Rutherford have to explain the results?

Rutherford reasoned that as some alpha particles appeared all over the screen, they must be hitting and 'bouncing off' something inside the atoms that repelled them, but, as most passed through, there must be a large amount of empty space in an atom to let the alpha particles through.

Electrons, protons and neutrons

After further thought, Rutherford concluded that an atom did not have a positively charged 'pudding' around the outside, but instead had a

18 How are the Thomson atomic model and the Rutherford atomic model
   a similar
   b different?

positively charged centre or **nucleus**, which was surrounded by negatively charged electrons. He used his thoughts to construct a new model of the atom, as shown in Figure 8.6.

In this model, he showed that he believed that the atom had a highly charged centre and was surrounded by a large amount of empty space, in which the electrons were spread out in an orderly way.

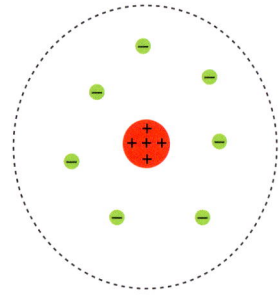

▲ **Figure 8.6** Rutherford's model of the atom.

Rutherford's later work identified particles in the nucleus which he named **protons** after Joseph Proust, who first suggested that atoms might have smaller particles inside them.

When Rutherford looked at all the evidence that had been collected about atomic structure, he found that there seemed to be something missing to explain all the results. He thought that there must be other particles present in the nucleus which were similar to protons but which did not have an electrical charge. He thought that for a structure to have no overall charge, it must be made of a positively charged proton and a negatively charged electron together, and he called the particle a 'neutral doublet'.

This critical look at the data led other scientists to evaluate the methods used for investigating atomic structure, refining them for further investigations. In 1932, an English scientist named James Chadwick (1891–1974) fired alpha particles at beryllium atoms and knocked out particles that had a similar mass to protons but no electrical charge. He had discovered the neutral doublet predicted by Rutherford, and he called it the **neutron**. Further work on this particle by others showed that it was not composed of a proton and an electron – rather, it was a particle with a similar structure to a proton but with no electrical charge.

19 What scientific knowledge and understanding did Rutherford use to describe the structure of the neutral doublet?

20 Assess the accuracy of Rutherford's prediction about a neutral particle.

The electrostatic charge inside the atom

We have seen in the models of Thomson and Rutherford that the nucleus had a positive electric charge and the electrons had negative electric charges. This difference in electrical charge creates an **electrostatic** force between the nucleus and the electrons in the atom. It is this electrostatic force, created by the attraction between positive protons and negative electrons, that keeps each atom together.

How can you demonstrate the electrostatic charge?

You will need:

an inflated balloon on a thread and a woollen jumper.

Process

Wear the woollen jumper and hold out the balloon from its thread. The balloon should hang down.

Rub the balloon on a sleeve. This will transfer electrons from the atoms in the wool to atoms in the rubber of the balloon, as Figure 8.7 shows.

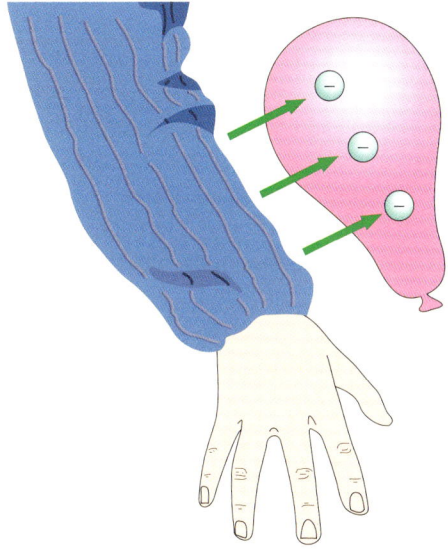

▲ Figure 8.7

The balloon is now negatively charged and the wool on the arm is positively charged.

Bring the balloon on its thread close to the sleeve. The differently charged objects have generated an electrostatic force between them.

Examining the results

What does this force do when you bring the balloon close to the sleeve?

Making a model

1 How could you use this demonstration to model the relationship between a proton and an electron?
2 In your model, what is the analogy for the:
   a nucleus
   b electron?

Viewing atoms

Atoms are too small to be seen by microscopes which use light, like the ones you may use in the laboratory. Scientists and engineers have developed microscopes which fire electrons at materials to make images of the atoms inside them. In the images, the detail of the atoms cannot be seen, but their position in the material can.

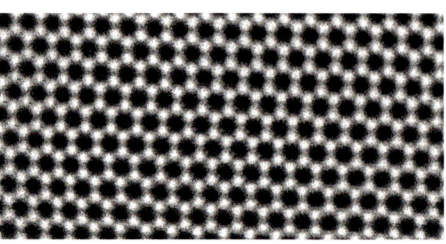

▲ **Figure 8.8** Each white spot in this hexagonal structure of graphene is a carbon atom.

21 Does the picture of atoms in a material support Democritus' idea? Explain your answer.

Summary

✔ Science in context: People have wondered about the structure of materials for a long time, and the term 'atom' was used by Democritus in ancient Greece to describe them.

✔ Science in context: Antoine Lavoisier investigated chemical reactions and established the law of the conservation of mass, a discovery built on by Joseph Proust.

✔ Science in context: John Dalton's atomic theory outlined the composition of matter and atoms, the properties of atoms of an element, and how they combine in reactions.

✔ Science in context: Joseph J Thomson devised the plum pudding model of atoms, with negatively charged electrons surrounded by a positively charged 'pudding'.

✔ Ernest Rutherford's experimentation with atomic structure led to his model of the structure of the atom.

✔ Electrons have negative charge, protons have positive charge and neutrons have no charge.

✔ The chemical properties of elements are linked to their atomic structure.

✔ Electrostatic force is the force created by the attraction between positive protons and negative electrons that keeps the atom together.

End of chapter questions

1   What creative thought did Democritus have that led to the idea of the atom?

2   What did Rutherford use to fire at gold to find out about its atoms?

3   What was the curved screen used for in Rutherford's experiment?

4   What did Rutherford find on the screen after his experiment?

5   What did Rutherford conclude from his experiment?

6   How are a proton and neutron
    a   similar
    b   different?

7   Compare the electrical charge of a proton with an electron.

8   What holds electrons inside an atom?

9   Look at the dates showing the lives of all the scientists whose work helped to build up ideas of atoms, and arrange their names into a timeline. Write up to two sentences about what each scientist in your timeline did.

 Now you have completed Chapter 8, you may like to try the Chapter 8 online knowledge test if you are using the Boost eBook.

9 Mixtures and impurities

In this chapter you will learn:
- that purity is a way to describe how much of a chemical is in a mixture
- about impurity and how impure mixtures can result from reactions
- about concentration of solutions and how they relate to solutes and solvents
- how the solubility of different salts varies with temperature
- about paper chromatography and how we can use this to identify substances and separate them from a sample
- about the uses of chromatography (Science in context).

Do you remember?

- What do you know about the properties of water?
- What happens when a solid dissolves in a liquid?
- What is a solvent?
- 'Substances can change state when their temperature rises or falls.' What does this statement mean?
- How is an element different to a compound?
- How is a compound different to a mixture?
- What is a mixture of metals called?

Pure and impure substances

Substances can be made of elements, compounds and **mixtures**. If a substance is made from atoms of just one kind of element, or from just one type of compound, then it is a pure substance.

Very few substances are pure. They usually contain other elements or compounds which are termed **impurities**. A substance containing impurities is described as an **impure** substance.

Modelling impure substances

Models can help us to form a picture of what happens in a mixture. We could represent this by drawing 2-D circles on a page. Create a model to show what happens when an impure substance is formed from
i) an element with a compound as an impurity, or ii) a compound with an element as an impurity. It could be a 2-D or 3-D model.

Detecting pure and impure substances

A pure substance has a definite **melting point** and **boiling point**. These are the temperatures at which the substance melts and boils. Experiments have been done to find the melting and boiling points of pure substances and the data collected has been recorded.

Comparing melting and boiling points of some common substances

You will need:

internet access.

Investigation and recording data

Use the internet to find the melting and boiling points of the following substances:

- acetic acid (vinegar)
- iodine
- sodium chloride (salt)
- pure gold

Examining the results

Check your data and look for a trend. If you find one, what is it?

Conclusion

Can you see a trend if you list the substances in order of melting and boiling points?

Causes of impurities

Sometimes reactants are present in such large amounts that not all of the reactant is used up in a reaction. You can test this with the following enquiry.

Is all the acid used up in a chemical reaction?

Work safely

Take care with hydrochloric acid and sodium hydroxide. They are irritants and they are corrosive.

Eye protection and safety gloves should be worn for this experiment.

You will need:

a test tube rack, a boiling tube containing 10 cm³ of dilute hydrochloric acid, a bottle of universal indicator solution and a dropper, a measuring cylinder, a beaker of dilute sodium hydroxide solution.

Hypothesis

Universal indicator solution produces a range of colours in acids and alkalis. If it is used to test the reaction between an acid and an alkali, it will turn green when all the acid has taken part in the reaction.

Prediction

When an alkali is added gradually to an acid containing universal indicator, it will eventually turn green, showing that the acid has been used up and a neutral solution has been obtained (i.e., pH 7, which gives a green colour).

Investigation and recording data

1 Set up the boiling tube in the test tube rack.
2 Add three or four drops of indicator solution to the acid in the boiling tube until a colour is clearly seen.
3 Pour out sodium hydroxide solution from the beaker into the measuring cylinder, up to the 5 cm³ mark on the scale.
4 Pour the sodium hydroxide solution from the measuring cylinder into the boiling tube of acid and make a note of the colour of the mixture.
5 Repeat steps 3 and 4 until the mixture starts to indicate that the neutralisation reaction has occurred – which would usually be identified by a colour change to green.

Examining the results

Use your knowledge of acids, alkalis, pH and universal indicator solution to explain the result.

Conclusion

Draw a conclusion. Does the examination of your results support the hypothesis?

What improvements could be made? Explain the changes that you suggest.

▲ **Figure 9.1** Himalayan rock salt.

- The temperature of the reactants may have been too low for all of them to completely react in a certain time. For example, copper oxide does not react with sulfuric acid at room temperature, and both reactants need to be warmed for the reaction to take place.
- A reactant may come to the reaction with impurities already present in it, such as rock salt, which contains minerals that colour the crystals.

CHALLENGE YOURSELF
Use balls of modelling clay or coloured beads to make two models:
a   a pure substance
b   a substance with an impurity.

Calculating purity

The purity of a product is expressed as a percentage. It is found by finding the mass of the pure substance in the sample of the product, and the mass of the whole sample. It is calculated with this equation:

% purity = mass of substance in the sample/mass of the sample × 100

CHALLENGE YOURSELF

1 A sample has a mass of 100 g but the mass of the substance in it is 60 g. What is the percentage purity of the sample?
2 Calculate the percentage purities of the following samples:
   a **Sample A** substance mass 35 g, sample mass 70 g
   b **Sample B** substance mass 65 g, sample mass 130 g
   c **Sample C** substance mass 27 g, sample mass 108 g

Dissolving and concentration

When a solid **dissolves** in a liquid it forms a **solution** with the liquid. The liquid in which the solid dissolves is called a **solvent**, and the solid which dissolves in the solvent is called the **solute**. The particle theory explains how things dissolve in the following way.

There are small gaps between the particles in a liquid. When a substance dissolves in a liquid, its particles spread out and fill in the gaps. Figure 9.2 shows how particles in a solid solute are pulled apart by the particles in the liquid solvent, which then move between them.

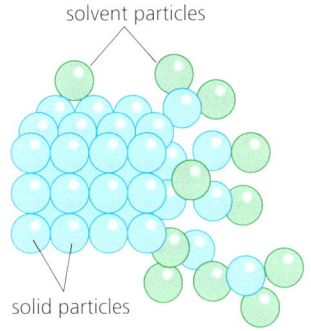

solvent particles

solid particles

▲ **Figure 9.2** Particles showing the dissolving process.

▲ **Figure 9.3** Copper sulfate dissolves in water to form a clear blue solution.

The term 'concentration' is used by scientists to describe the amount of solid particles that are dissolved in a volume of solvent. If a small number of solid particles dissolve in a volume of solvent, the concentration is said to be low, as shown in Figure 9.4a on the next page. If a large number of solid particles dissolve in a volume of solvent, the concentration is said to be high, as shown in Figure 9.4b. If the solid makes a colourless solution when it dissolves, you cannot tell its concentration.

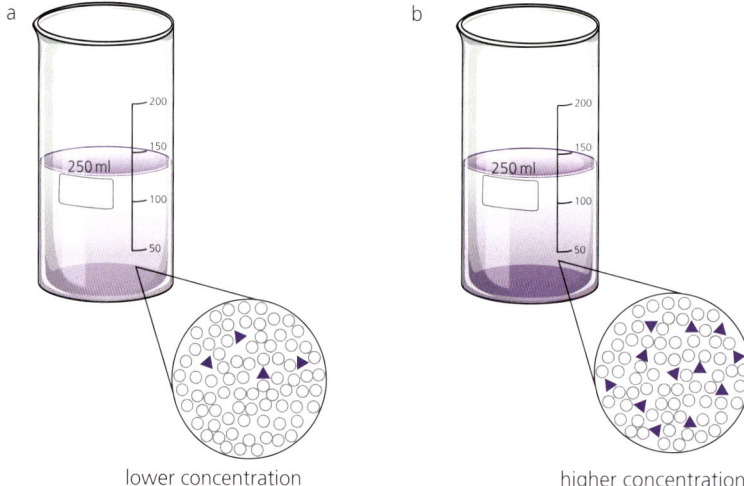

▲ **Figure 9.4a** A magnified view of a beaker showing low concentration of solid particles; **b** a magnified view of a beaker showing a higher concentration of solid particles.

CHALLENGE YOURSELF

Can you assess the concentration of a solution by its colour?

Put 50 cm³ of water in a beaker with four drops of food colouring. Photograph the solution. Now add another 50 cm³ of water, stir with a stirring rod and photograph again. Repeat twice more until you have 200 cm³ in the beaker. Use your observations to answer the question.

1 Which substance is more soluble at 0 °C?

2 What is the temperature when both substances are equally soluble?

3 What is the solubility of potassium nitrate at 40 °C and 60 °C?

4 What does the graph tell you about the solubilities of sodium chloride and potassium nitrate?

Solubility

Solubility is a property of every substance. In simple terms, it is the ability of a substance to dissolve in a particular solvent. In scientific terms, it is described as the maximum amout of a substance that will dissolve in a certain amount of solvent at a certain temperature. Different substances have different solubilities. Potassium nitrate and sodium chloride are both compounds known as salts. They have very different solubilities at different temperatures, as Figure 9.5 shows.

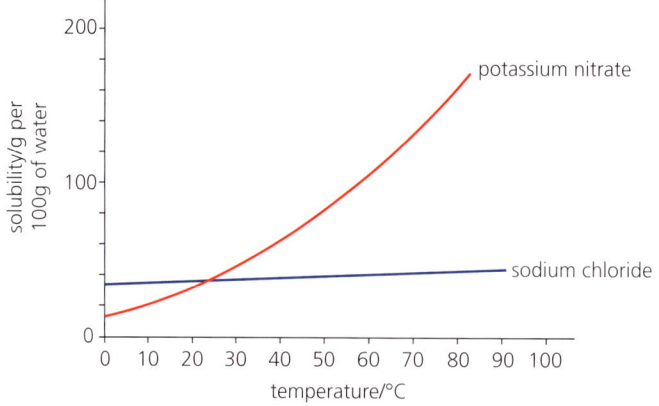

▲ **Figure 9.5** The effect of temperature on the solubility of potassium nitrate and sodium chloride.

5 Which substance in Table 9.1 is most soluble at 0°C?

6 Which substance in Table 9.1 is most soluble at 100°C?

7 Which substance in Table 9.1 shows the least increase in solubility at 100°C?

8 Which substance in Table 9.1 shows the greatest increase in solubility at 100°C?

Table 9.1 shows how the solubility of some other compounds changes with a rise in temperature. The numbers in the table show the solubility of the substance in grams per 100 ml at 0°C and 100°C.

Substance	0°C	100°C
Copper sulfate	23.1	114
Magnesium chloride	52.9	73.3
Potassium carbonate	105	156
Silver nitrate	122	733
Zinc chloride	342	614

▲ Table 9.1

Why solubility increases with temperature

Look back at Figure 9.4 and see how the solvent particles are surrounding the solid particles. When the temperature rises, the **energy** of all the particles rises. The solid particles vibrate more and the solvent particles move faster in all directions, hitting the group of solid particles more frequently. This increased movement of particles due to the increase in temperature causes the solid particles to separate and spread out between the solvent particles.

Chromatography

A solvent may have two or more solutes dissolved in it. One way to check the number of solutes in a solvent is to use a process called **chromatography**. The stages in this process are set out in Figure 9.6.

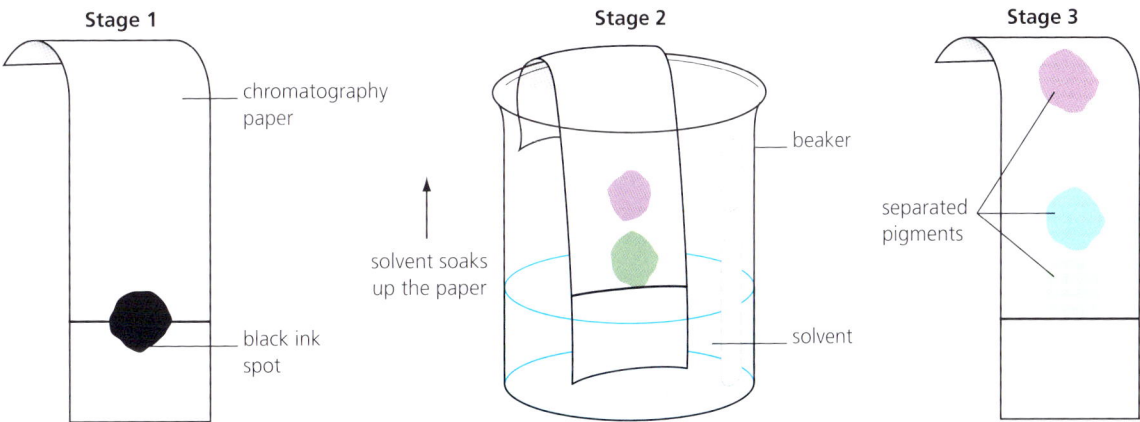

▲ **Figure 9.6** Simple paper chromatography.

Stage 1

A pencil line is drawn across a strip of chromatography paper near one end. The pencil mark does not dissolve in the solvent. A drop of liquid is placed on the pencil line.

Stage 2

A solvent is placed in a beaker. It should only reach a depth which is below the level of the pencil mark. The solvent which is chosen must be one in which the solutes (in the sample) are soluble. If they are not soluble in the solvent, they will not move up the chromatography paper. Once the correct solvent has been selected and poured into the beaker, the end of the chromatography strip (with the sample on it) should be dipped into the beaker and left for a while.

In this stage, the chromatography paper absorbs the solvent and draws it up the strip. The solutes in the sample dissolve in the solvent and move up the paper with it. Each type of solute moves at a different speed to the others, so they spread out up the paper. They do this because the pigments vary in their solubility and in their tendency to stick to the paper. The most soluble pigments with the least tendency to stick to the paper move furthest up the paper, while the least soluble pigments with the most tendency to stick to the paper move the least distance from the pencil mark.

Stage 3

When the solvent has risen almost to the top of the paper, it can be removed and dried, and the separated solute can be observed. Each solute will move up the paper to the same position, no matter which sample it is in, if the same solvent is used. This means that if two different samples are tested and have a solute which moves the same distance up each strip, it is the same type of solute that is present in both samples.

> **DID YOU KNOW?**
> The paper with its separated solutes is called a **chromatogram**.

Do different inks have the same solutes?

You will need:

strips of chromatography paper, beakers, water-based inks or felt-tip pens containing water-based inks, and water.

Plan, investigation and recording data
Look at the stages in a chromatography experiment, as shown in Figure 9.6, and work out a plan to investigate a range of inks. If approved by your teacher, try it.

Examining the results
Examine the dried strips of chromatography paper. Look for solutes in the same places on different strips.

Conclusion

Draw a conclusion. Do your results agree or disagree with the enquiry question? Explain your answer.

Is your conclusion limited in some way? Explain your answer.

What improvements could be made? Explain the changes that you suggest.

Chromatography has a wide range of uses. Here are some examples.

Science in context

The uses of chromatography

Foods are made from many substances, such as nutrients like protein, but they also contain substances which can break down and produce acids, which spoil the taste of the food and make it uneatable. A type of chromatography called **column chromatography** is used to check foods to see that they are not about to spoil before they are sold. Some foods have substances called **additives** added to them to make them taste better. The additives have certain chemicals in them that separate during chromatography, which shows whether the food contains its natural flavourings or has had additives added.

▲ **Figure 9.7** Column chromatography.

In forensic science, a form of chromatography has been developed called **gas chromatography**. The substances to be investigated are allowed to pass through a tube of gas inside the equipment, and they settle out at different distances to make them easier to identify. This type of chromatography is used to test hair and blood samples found at the scene of a crime.

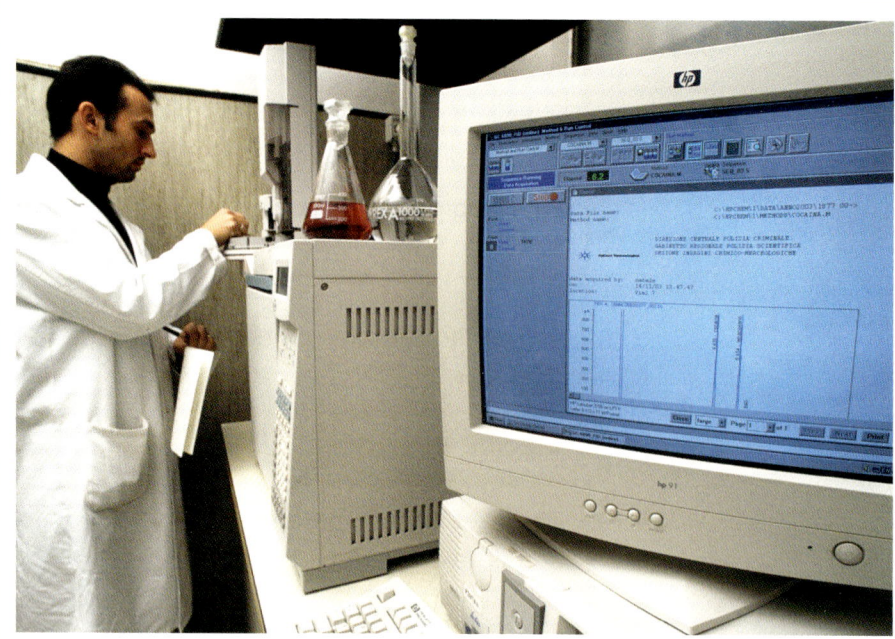

▲ **Figure 9.8** The scientist is preparing a sample for gas chromatography. The separation of the substances is shown on the screen.

Chromatography is also used to test for the presence of chemicals in the environment, such as pesticides, to assess pollution and the danger to ecosystems. It is also is used in medicine in the preparation of drugs, to check for purity, and in the development of **vaccines** – substances which protect people from developing diseases.

LET'S TALK

Imagine you were a food producer and you received a report from the chromatography laboratory saying that your food showed signs of being spoilt. What would you do?

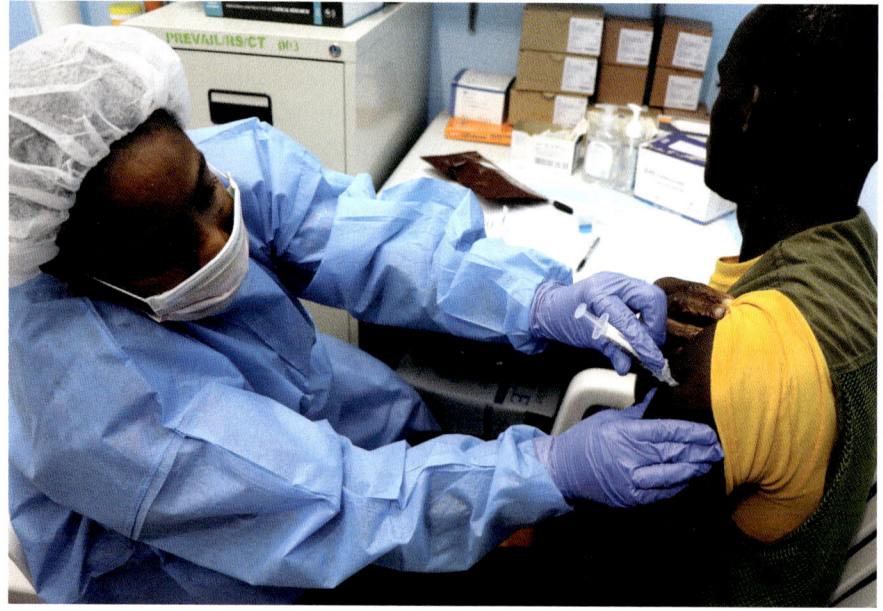

▲ **Figure 9.9** A vaccine for Ebola being trialled in west Africa.

Summary

✔ Purity is a way to describe how much of a chemical is in a mixture.
✔ A substance that contains atoms from only one element or type of compound is pure.
✔ An impure substance is one which contains other elements or compounds, called impurities.
✔ Impure mixtures can result from reactions where the temperature is too low, the amount of one reactant is too high, or where there are impurities already present.
✔ Concentration describes the amount of solid particles that are dissolved in a volume of solvent.
✔ Paper chromatography can be used to separate and identify substances in a sample and can be used to test food, for pesticides and pollution, as well as in medicine and vaccine preparation.

End of chapter questions

1 What is a pure substance?
2 What is an impurity?
3 Two reactants produced a product, but when it was tested for purity, it still had one of the reactants in it. How could this have happened?
4 Why should you make sure that a chemical reaction takes place at the correct temperature for the correct amount of time?
5 A solution has a high concentration of salt in it. What does this mean?
6 Imagine you have a glass of orange juice, and you pour it into a larger glass with a glass of water.
   a What happens to the colour of the orange juice?
   b Explain why the orange juice changed colour.

 Now you have completed Chapter 9, you may like to try the Chapter 9 online knowledge test if you are using the Boost eBook.

10 The reactivity of metals

In this chapter you will learn:
- about the reactivity of metals with oxygen, water and dilute acids
- how to use word equations to describe reactions
- that some substances are generally unreactive and can be described as inert.

Do you remember?

- What are substances that take part in a chemical reaction called?
- What happens to these substances in a chemical reaction? Do their amounts increase or decrease?
- What are new substances which are formed in a chemical reaction called?
- When substances are brought together, how can you tell if a chemical reaction is taking place?

In a chemical reaction, new substances are formed. The **reactivity** of a substance is a description of how readily it takes part in chemical reactions with other substances.

Sodium is so reactive that it has to be kept under a layer of oil. The reason for this is that it will readily react with water and oxygen in the air and produce flames. Sodium is a very soft metal and can be cut with a knife. If a small piece is cut and exposed to the air, its shiny surface quickly tarnishes as it reacts with the air.

When iron reacts with damp air, a chemical reaction takes place and a new substance called **rust** is formed. This reaction takes place much more slowly than the reaction between sodium and the air, which suggests that some metals, such as sodium, react more vigorously than other metals, such as iron, in chemical reactions.

▲ **Figure 10.1a** Sodium reacts so strongly with oxygen and water that it must be stored under oil to stop it reacting; **b** Iron reacts more slowly, but over time the results can be dramatic.

Reaction of metals with oxygen

Here are some descriptions of the reactions that take place when certain metals are heated with oxygen.

- Copper develops a covering of black powder without glowing or bursting into flame.
- Thin iron wire, called iron wool, glows and produces yellow sparks; a black powder is left behind.
- Sodium only needs a little heat to make it burst into yellow flames (Figure 10.2a) and burn quickly to produce a new substance in the form of a white powder (Figure 10.2b).
- Gold is not changed after it has been heated and then left to cool.

▲ **Figure 10.2a** Sodium burns rapidly in a gas jar of oxygen; **b** Sodium oxide powder is left behind.

1 Arrange the metals mentioned in this section in order of their reactivity with oxygen. Start with what you consider to be the most reactive metal.

2 Write word equations for the metals that reacted with oxygen in this section.

The general word equation for a metal that reacts with oxygen is:

metal + oxygen → metal oxide

The example of sodium in Figure 10.2 (above) can thus be represented by the word equation:

sodium + oxygen → sodium oxide

Sodium oxide is the product of the chemical reaction.

Reaction of metals with water

Here are some observations that can be made when a metal is placed in water.

- Calcium sinks in cold water and bubbles of hydrogen form on its surface, slowly at first. The bubbles quickly increase in number and the water becomes cloudy as calcium hydroxide forms. The bubbles of gas can be collected by placing a test tube filled with water over the fizzing metal. The gas pushes the water out of the test tube. If the tube, now filled with gas, is quickly raised out of the water and a lighted splint is held beneath its mouth, a popping sound is heard. The hydrogen in the tube combines with oxygen in the air and this explosive reaction makes the popping sound.
- Copper sinks in cold water and does not react with it.
- Sodium floats on the surface of the water and a fizzing sound is heard as bubbles of hydrogen are quickly produced around it. The production of the gas may push the metal across the water's surface and against the side of the container, where the metal bursts into flame. A clear solution of sodium hydroxide forms.
- Iron sinks in water and no bubbles of hydrogen form.
- Magnesium sinks in water. Bubbles of hydrogen are produced only very slowly and a solution of magnesium hydroxide is formed.
- Potassium floats on water and bursts into flames immediately (Figure 10.3). Hydrogen bubbles are rapidly produced around the metal. A clear solution of potassium hydroxide forms.

▶ **Figure 10.3** The reaction of potassium with water gives out enough heat to set alight the hydrogen produced, which burns with a pink flame because potassium is present. The heat also melts the potassium metal, which forms a molten ball that floats on the water's surface.

3 Using the information in this section, arrange the metals in this section in order of their reactivity with water.

4 In some houses, copper is used to make hot water tanks, while steel (a modified form of iron) is used to make cold water tanks. Why can steel not be used to make the hot water tank?

Water is a compound of hydrogen and oxygen. In a chemical reaction with a metal, the compound is broken up and forms a metal hydroxide and hydrogen as the word equation shows.

metal + water → metal hydroxide + hydrogen

5 Write word equations for the metals that reacted with water in this section.

You can check some of the previous observations about the metals and water with the following scientific enquiry. The indicator in this enquiry is phenolphthalein. It changes from colourless to pink when a metal has reacted with water to produce the metal hydroxide.

DID YOU KNOW?

When iron reacts with water and oxygen in the air, it forms rust and a substance made from a mixture of two different iron oxides. They form brown flakes which fall off the metal and expose more of it to the water and oxygen around it.

▲ **Figure 10.4** A rusty iron gate.

Do all metals react with water in exactly the same way?

You will need:

four boiling tubes, small samples of zinc, magnesium, iron and copper, a 600 cm³ beaker, a bottle of the indicator phenolphthalein with a dropper, a stirrer and some water.

Investigation and recording data

1 Half-fill the beaker with water.
2 Fill the dropper with indicator and squirt it carefully into the water in the beaker.
3 Stir the water and indicator together.
4 Place a sample of each metal in a separate boiling tube.
5 Half-fill each boiling tube with the water and indicator mixture.
6 Observe the samples in the tubes over the next few minutes and look for a pink colour in any tube. If you see a pink colour it indicates that the metal has reacted with the water.

Examining the results
Examine the contents of the four boiling tubes after a few minutes.

Conclusion
Draw a conclusion after examining your results.

Reaction of metals with acids

Sodium and potassium react violently with acids, and these reactions should not be carried out. Other metals can be tested for their reaction using a beaker, as shown in Figure 10.5, and using the equipment in Figure 10.6 to collect the hydrogen gas for testing with a lit splint.

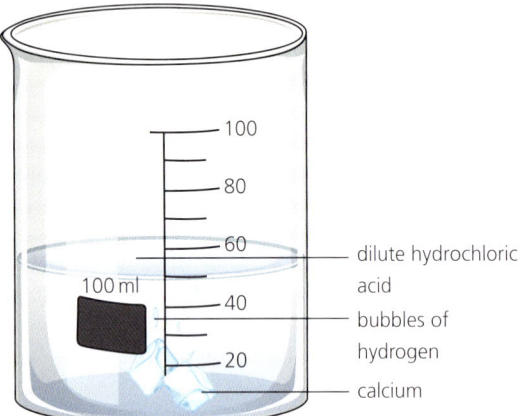

▲ **Figure 10.5** Testing how calcium reacts with hydrochloric acid.

dilute hydrochloric acid

bubbles of hydrogen

calcium

100 ml

100
80
60
40
20

▲ **Figure 10.6** Equipment used in investigating the reaction of some metals (not sodium or potassium) and hydrochloric acid.

A piece of metal is placed in a test tube or flask, then acid is poured in and the bubbles of hydrogen gas are observed. The reactivity of metals can be compared by simply comparing the amount of hydrogen gas produced in the reaction between the acid and the different metals. The general word equation for this reaction is:

metal + hydrochloric acid → metal chloride + hydrogen

The equipment shown in Figure 10.6 is designed to collect any hydrogen that is produced in the reaction. When the acid is poured in down the thistle funnel, it should reach a level above the bottom of the funnel tube. The hydrogen escapes from bubbles at the surface of the solution and pushes air out of the flask and down the delivery tube. One minute should be allowed for the air to escape from the end of the delivery tube, and then a test tube full of water can be put over it to collect the hydrogen. As the gas collects in the upturned test tube, it pushes water out at the bottom. When the tube is full of gas, it can be closed with a bung or a stopper and replaced with another tube to collect any more gas that is produced. The amount of hydrogen produced will depend on the amount of metal (the mass of metal) and the concentration of the acid.

6 Based on the description of the way sodium and potassium react with water, which metal do you think will react more violently with acids?

7 What might happen to the hydrogen if the bottom of the funnel in Figure 10.6 was above the surface of the liquid?

8 What is the word equation for the reaction of each of the following metals with hydrochloric acid?
   a magnesium
   b iron
   c zinc

Here are some descriptions of the reactions that take place between different metals and hydrochloric acid. If scientists cannot get a reaction with a dilute acid, they test with concentrated acid. You must never use concentrated acid – only acids approved by your teacher.

- Lead did not react with dilute hydrochloric acid, but when tested with concentrated acid, bubbles of hydrogen gas were produced slowly.
- Zinc reacted quite slowly with dilute hydrochloric acid to produce bubbles of hydrogen.
- Copper did not react with either dilute or concentrated hydrochloric acid.
- Magnesium reacted quickly with dilute hydrochloric acid to produce bubbles of hydrogen.

9 Arrange the metals in this section in order of their reactivity with hydrochloric acid.

You can check some of the information given on the pages of this section by trying the following scientific enquiry.

Do all metals react in the same way with acids?

Work safely

Make sure that the equipment has cooled down before clearing it away.

If carrying a burning splint from a Bunsen burner or spirit burner that is already lit, take great care not to bump into anyone.

You will need:

a test tube rack, a bung, test tubes, a boiling tube, some splints, a Bunsen burner or spirit burner, a heat-proof mat, 1M hydrochloric acid, samples of magnesium, calcium, iron and zinc, safety glasses and a device for recording sound (optional).

Read through the following investigation and discuss it with your group and teacher to identify and resolve any safety issues.

Investigation and recording data

1 Light the Bunsen or spirit burner and keep it on a safe flame.
2 Put a sample of magnesium in a test tube in the rack.
3 Half-fill the test tube with the hydrochloric acid.
4 When the bubbles start to appear, invert the boiling tube over the test tube. Any hydrogen released from the bubbles will rise into the boiling tube and push the air out.
5 Raise the inverted boiling tube carefully above the test tube and insert a bung into its mouth (opening) so that it stays in place, but do not push too hard.
6 Turn the boiling tube the right way up and place it in a boiling tube rack.
7 Bring a lighted splint close to the boiling tube, carefully remove the bung and move the splint to the mouth of the boiling tube.
8 Listen for a squeaky pop that indicates the presence of hydrogen. You may record the sound if you wish.
9 Repeat steps 2–8 with each of the other metals.

Examining the results

Examine your data and compare the reactions of the metals.

Conclusion

Draw a conclusion after examining your results.

The reactivity series

Chemists have taken the results of the three tests with oxygen, water and acids and set up a **reactivity series** with the most reactive metals at the top of the series and the least reactive metals at the bottom, as shown in Table 10.1.

Substances which do not react with other substances are called **unreactive** substances. These unreactive substances are also sometimes described as being **inert**.

▼ **Table 10.1** The reactivity series and the performance of the metals with oxygen, water and acids. When scientists found a metal was unreactive with cold water, they tested it by heating it in steam. You must never try this test.

Metal	Reaction with oxygen	Reaction with water	Reaction with acid
potassium	oxide forms very vigorously	produces hydrogen with cold water	violent reaction
sodium			
calcium		produces hydrogen with steam	rate of reaction decreases down the table
magnesium			
zinc			
iron	oxide forms slowly		
copper	oxide forms without burning	no reaction with water or steam	no reaction
silver	no reaction		
gold			

10 Using Table 10.1 which substances in the reactivity series can be described as inert? Explain your answer.

Summary

✔ In a summary word equation, the reactants are shown on the left of the arrow and the products are shown on the right of the arrow.
✔ The reaction of a metal with oxygen creates a metal oxide.
✔ The reaction of a metal with water creates a metal hydroxide and hydrogen.
✔ The reaction of some metals with a hydrochloric acid creates a metal chloride and hydrogen.
✔ The reactivity series shows the most reactive metals at the top of the series and the least reactive metals at the bottom.
✔ Some substances are generally unreactive and can be described as inert.

End of chapter questions

1 Why is sodium kept under oil?
2 What happens to copper when it is heated in oxygen?
3 What happens to gold when it is heated in oxygen?
4 If a metal reacts with oxygen, what is the product?
5 If a metal reacts with water, what is the product?
6 Name two metals that float on water.
7 What is phenolphthalein and what is it used for when testing metals and water?
8 What is the purpose of the reactivity series and what tests are used to set it up?
9 What is rust and how does it form?

 Now you have completed Chapter 10, you may like to try the Chapter 10 online knowledge test if you are using the Boost eBook.

11 Exothermic and endothermic reactions

In this chapter you will learn:

- about exothermic reactions
- about measuring energy in fuel and foods (Science in context)
- about improving fuel efficiency (Science in context)
- about hand warmers and rusting (Science in context)
- about endothermic reactions
- further word equations to describe reactions
- that exothermic and endothermic reactions can be identified by temperature change
- the chemistry of ice packs (Science in context).

Do you remember?

- When something changes its state, what does it do?
- What happens to something when it freezes?
- Explain what happens when something melts, using the particle model to help you.
- When something changes state, is it a physical process or due to a chemical reaction?
- How can you tell if a chemical reaction has taken place?

▲ **Figure 11.1** A campfire gives out heat (an exothermic reaction).

An **exothermic reaction** is chemical reaction in which energy is transferred to the surroundings in the form of heat. An **endothermic reaction** is a chemical reaction in which energy is transferred from the surroundings in the form of heat. Both these reactions can be detected by a change in temperature.

Exothermic reactions

Burning

The exothermic reaction that is most familiar to everyone is **combustion**. When a flame develops in this reaction, combustion is then called **burning**. We say that a burning substance is on fire.

◀ **Figure 11.2** The burning oil in these lamps gives out light.

1 Give a use for each of the fuels listed in the paragraph on burning. How many different uses can you find?

Many substances are burned to provide heat or light. These substances are called **fuels**. Wood, coal, coke, charcoal, oil, diesel oil, petrol, natural gas and wax are all examples of fuels. The heat may be used to warm buildings, cook meals, make chemicals in industry, expand gases in vehicle engines and turn water into **steam** to drive generators in power stations. Some gases, oils and waxes are burned in lamps to provide light.

Measuring the energy in a fuel

The energy in a fuel can be found by collecting the heat energy released from a burning fuel and measuring it. This can be done by working scientifically in the following way.

The mass of the fuel is measured before being burnt, and the temperature of the water in a beaker is taken before it is heated by the fuel. During the heating process, the temperature of the water in the beaker is observed, using the thermometer, until it has risen by about 10 °C; then the burner is removed and the flame extinguished, and the mass of the fuel is measured again. All measurements should be precise. They are then used to perform a calculation.

Can the energy in fuel be investigated simply?

Work safely

Make sure that the equipment has cooled down before clearing it away.

If carrying a burning splint from a Bunsen burner or spirit burner that is already lit, take great care not to bump into anyone.

You will need:

a beaker, a thermometer, a tripod, gauze, a spirit burner and a top-pan balance or other weighing machine.

Investigation and recording data

1 Pour a known volume of cold water into the beaker and record its temperature.
2 Weigh the spirit burner containing the fuel and record its mass.
3 Light the burner so that the burning fuel heats the water.
4 Take the temperature of the water.
5 When the temperature has risen by 10 °C, put out the burner flame, record the temperature of the water, and then reweigh the burner and fuel.

Examining the results

Find the amount of fuel used by subtracting the mass of the burner and fuel *after* the experiment from the mass of the burner and fuel *before* the experiment.

Find the rise in temperature by subtracting the temperature of the water at the *beginning* of the experiment from the temperature of the water at the *end* of the experiment.

Conclusion

Draw a conclusion. Can the energy transferred when a fuel burns be measured in some way? Use the results of this experiment to explain your answer.

Is your conclusion limited in some way? Explain your answer.

What improvements could be made? Explain the changes that you suggest.

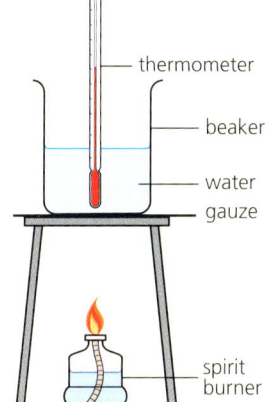

▲ **Figure 11.3** The arrangement of equipment for the investigation.

CHALLENGE YOURSELF

How could you use the procedure with the beaker of water to compare the energy released by different fuels? Work out a plan and, if your teacher approves, try it.

Science in context

Measuring energy in fuel and foods

When you have made an investigation, you are often asked how you could make improvements. When scientists discovered that they could heat water to find out the energy in a fuel, using equipment as shown in Figure 11.3, they looked for ways to improve the equipment to measure the energy in foods. They used creative thinking to improve on the equipment, as shown in Figure 11.4.

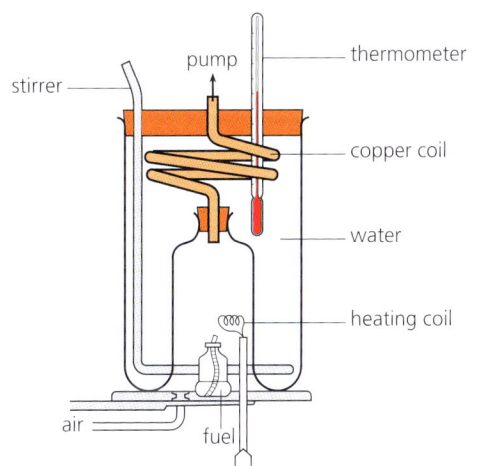

◀ **Figure 11.4** A specially designed calorimeter.

2 Why is air pumped through the chamber instead of just letting the fuel use the air that is present?

3 The pipe carrying the hot gases is made of a copper coil. What do you think is the reason for this?

4 Why is the water stirred?

5 How do you find the rise in temperature of the water?

The fuel was placed in a specially built glass flask which had water around it. The temperature of the water was taken, then an electric current was passed though the heating coil to light the burner. A pump was switched on to draw air through the flask as the fuel burnt. The hot air produced as the fuel burnt passed through the copper coil.

Copper is a very good conductor of heat, so this helped to transfer the heat to the water. The copper pipe was shaped into a coil so that a longer length could be enclosed in the water. This was done to let the hot air travel a greater distance and give out more heat to the pipe and water.

This set of equipment is known as a bomb calorimeter. Today, even more complicated bomb calorimeters are used to measure the energy in samples of food by burning them in oxygen. This is how the energy values of the fuels and foods in Tables 4.2–4.5 (pages 32–36) were measured.

CHALLENGE YOURSELF

Table 11.1 shows the energy in one gram of a variety of fuels. The energy is measured in a unit called kilojoules, which has the symbol kJ. Use the table to answer the following question.

▼ **Table 11.1** Energy in various fuels.

Fuel	Energy/kJ per g
natural gas	56
oil	48
coal	34
coke	32
wood	22
carbohydrates (eg, starch and sugars)	16

Bjorn and Ingrid live in a country with cold winters. They live in identical houses, but Bjorn heats his house with coal, while Ingrid heats her house with wood. They both fill up their fuel stores for the winter. Who will need the larger store? Explain your answer.

Science in context

Improving fuel efficiency

The burning of a fuel depends on the oxygen in the air around it. If air can pass over the fuel freely, a large amount of the oxygen in it can take part in the combustion process, and a great deal of heat can be released. If the amount of air reaching the fuel is reduced, the amount of oxygen taking part in combustion is reduced, and the amount of heat released is also reduced. The control of heat is important in cooking, and can be controlled simply, as the design of the two-pot cooking stove shows (Figure 11.5).

Wood was the first fuel. In many parts of the world it is still used as a fuel today. It is often in short supply, so ways have been developed to use the fuel more efficiently. Figure 11.5 shows a stove that has been developed in Sri Lanka to provide heat for cooking two pots at once. One pot is used to boil rice while the other is used to cook vegetables.

The damper can be raised or lowered in order to control the amount of air reaching the fire. This in turn controls the burning of the wood. The baffle is used to control the direction of the flames. They can be made to go straight up in order to heat the pot above them.

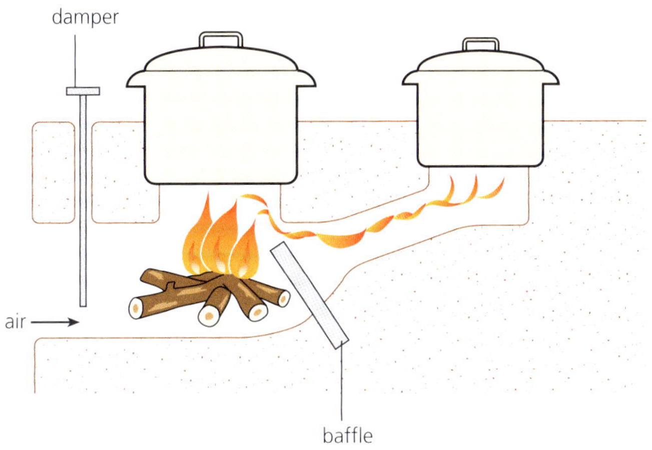

damper

air

baffle

◀ **Figure 11.5**
Heating two pots is more efficient than heating just one with the same amount of fuel.

6 How will designing more efficient stoves help to conserve fuel?

7 How will using the damper make the wood burn more slowly or more quickly?

8 In the past, people used to put a pan on three stones over a fire. Why is the stove in Figure 11.5 an improvement?

DID YOU KNOW?

A process called nuclear fission, in which a large atom splits up to form smaller atoms, is another exothermic reaction. This process is used in nuclear power stations to generate heat, which in turn is used to make electricity.

CHALLENGE YOURSELF

Is respiration an exothermic reaction? Look back at Chapter 3 (page 15) to find an answer, and then explain it.

Science in context

Hand warmers and rusting

Rusting is an exothermic reaction, but as the reaction is slow, the heat is produced in small amounts and quickly spreads into the air so that a rusting object does not feel warm. Useful heat can be produced from rusting in the design of some kinds of hand warmers. Iron powder is mixed with particles of charcoal (carbon), salt water, and an insulating substance such as vermiculite. All these constituents of the hand warmer are enclosed in a sealed airtight package. When the hand warmer is required, the seal is broken and air mixes with the constituents, so that an **oxidation** reaction takes place between the iron powder and oxygen. The carbon particles take up the heat and spread it out through the package, while the insulating material prevents it from escaping too quickly, so that a smaller amount of heat is released steadily to warm the hands.

▲ **Figure 11.6** An example of a handwarmer.

Endothermic reactions

Melting

Probably the most familiar endothermic process is melting. Once the temperature of the surroundings rises above 0 °C, ice absorbs heat energy from the air and starts to melt. This is a reversible physical reaction, not a chemical reaction.

Sherbet

Sherbet is a popular sweet, made from citric acid and sodium hydrogencarbonate. When you put sherbet in your mouth, it feels cool. This is due to an endothermic reaction taking place which results in heat being taken from your body. The reaction occurs when the sherbet dissolves in your saliva and the two chemicals react together. The word equation for the reaction is:

citric acid + sodium hydrogencarbonate → sodium citrate + carbon dioxide + water

Checking an endothermic reaction

An endothermic reaction is a chemical reaction in which energy is transferred *from* the surroundings. When this reaction takes place, energy in the form of heat moves from the surroundings. This means that the temperature, which is a measure of heat energy, falls. To record this fall in temperature, the reactants need to be enclosed in a container which has insulating walls. When citric acid and sodium hydrogen carbonate are mixed together they do not react chemically. They need water to do so. As soon as water is added, the endothermic reaction begins. Think about this information as you plan the following enquiry.

▲ **Figure 11.7** The sherbet inside these sweets makes your mouth feel cool.

9 What makes the sherbet fizz?

Can a fall in temperature be detected in an endothermic reaction?

You have available:

a thermometer or a data logger, a plastic cup, a metal cup, a thick-walled (styrofoam) plastic cup, a spatula, a measuring cylinder, citric acid powder, sodium hydrogencarbonate powder, some water, a stop-watch and access to a top-pan balance.

Plan, investigation and recording data

Select items from the list to use in your experiment.

Write a plan in which you record the amount of chemicals taking part in the reaction. Set out the steps in your experiment clearly, including how you will record your data, and show them to your teacher. If your teacher approves, try your experiment.

Examining the results

Examine the data that you have collected. Decide if the experiment needs modifying and repeating and, if so, check your changes with your teacher for approval.

Conclusion

Draw a conclusion which answers the enquiry question asked at the start of this box.

Is your conclusion limited in some way? Explain your answer.

What improvements could be made? Explain the changes that you suggest.

LET'S TALK

What changes take place in your favourite foods when they are cooked? How good are you at controlling the heat entering food (cooking) to make a meal that is neither uncooked nor burnt?

Cooking

When foods are cooked, the heat energy they take in allows chemical reactions to take place which change their structure and taste.

▲ **Figure 11.8** Cooking brings about chemical reactions that change food.

Science in context

The chemistry of ice packs

The instant ice pack is used to treat sports injuries. Inside the pack are two bags. One bag contains a chemical compound that contains nitrogen. It could be ammonium nitrate, calcium ammonium nitrate or carbamide. The second bag is inside the first and it contains water. When the ice pack is needed, it is squeezed. This breaks the bag of water in the centre, and the water mixes with the nitrate salt in a chemical reaction. This reaction draws in heat from the surroundings, making the ice pack cooler. If the ice pack is applied to the injury, it removes heat from that part of the body.

This reduces any swelling that might have occurred, reduces bleeding and eases pain by reducing activity in the nerves in the area, slowing the electrical signals that are sent to the brain.

▲ **Figuer 11.9** An ice pack being used for a sports injury.

Can you find the most useful chemical for an ice pack?

You will need:

sodium bicarbonate, potassium chloride, citric acid, sodium carbonate, calcium chloride, a beaker, a measuring cylinder, and a thermometer.

Which of these chemicals may be most useful in an ice pack?

Plan, investigation and recording data

Construct a fair test to compare how each chemical reduces the temperature of the water when mixed with it in turn. You may wish to look at the information in the previous *Science in context* box to work out your plan. Note that you do not need bags.

In your plan, set up a table in which to record your results. If your teacher approves your plan, try it.

Examining the results

Examine the data in your table and make comparisons between the different chemicals.

Conclusion

Use your evaluation to provide an answer to the enquiry question.

Is your conclusion limited in some way? Explain your answer.

What improvements could be made? Explain the changes that you suggest.

Summary

✔ An exothermic reaction is a chemical reaction in which energy is transferred to the surroundings in the form of heat.
✔ We can find the energy in fuel or foods by measuring the mass lost by burning in raising the temperature of water by 10°C.
✔ An endothermic reaction is a chemical reaction in which energy is transferred from the surroundings in the form of heat.

End of chapter questions

1 If something is on fire, what chemical reaction is taking place?
2 State three examples of a fuel.
3 What are fuels used for?
4 How do you use a beaker of cold water and a thermometer in an experiment to show that burning fuel is an exothermic reaction?
5 What is the exothermic reaction that is taking place inside our bodies to keep us warm?
6 When you have sherbet in your mouth, why does it feel cool?
7 What kind of reaction occurs in
   a a fuel when it is used to cook food
   b the food as it cooks?
8 Explain what happens when an ice pack is squeezed, and how it helps when someone has an injury.

 Now you have completed Chapter 11, you may like to try the Chapter 11 online knowledge test if you are using the Boost eBook.

In this chapter you will learn:
- how to calculate speed by knowing that speed = distance ÷ time
- about ways to measure speed (Science in context)
- how to interpret and draw simple distance/time graphs.

Do you remember?

- The distance between two places is measured in units of length. What units of length do you know?
- What are the units we use to measure time?
- What do you understand by the word 'speed'?

Some examples of speed

We all walk at a low **speed**, but when we run our speed is higher. Some people enjoy moving as fast as they can and use machines to help them do it, but they must do it safely for themselves and others. A few people aim to be the fastest movers over land on the planet. They compete with others for the world land speed record.

▲ Figure 12.1

▲ **Figure 12.2** Jogging is a form of exercise.

▲ **Figure 12.3** A fully kitted motorcyclist on a circuit.

The speed equation

Speed is a measure of the distance covered by an object in a certain time. It can be represented by this equation:

$$\text{speed} = \frac{\text{distance}}{\text{time}}$$

1 Write down an explanation of the speed equation, using the words 'equals' and 'divided by'.

Speed records

People trying to beat the land speed record must drive their vehicle at full speed between two markers.

▶ **Figure 12.4** Breaking the land speed record in 1997.

Table 12.1 shows some land speed records from the end of the twentieth century.

▼ **Table 12.1** Land speed records.

Date	Speed/kmph	Driver	Vehicle
15 October 1997	1223.66	Andy Green	Thrust SSC
4 October 1983	1020.41	Richard Noble	Thrust 2
23 October 1970	1014.66	Gary Gabelich	The Blue Flame
15 November 1965	955.95	Craig Breedlove	Spirit of America
7 November 1965	927.87	Art Arfons	Green Monster

> **CHALLENGE YOURSELF**
>
> Use the internet to see if the 1997 speed record has been broken. If it has, work out how much faster the vehicle went than the one driven by Andy Green.

2 For how long did Art Arfons' record stand?

3 How much faster than Spirit of America was The Blue Flame?

4 By how much did the land speed record rise between 15 November 1965 and 15 October 1997?

Investigating speed

There are many investigations that you can make about speed. Here are some of them.

Can you collect speed data and draw a conclusion from it?

You will need:

a clamp and stand, a toy car, a 1m long wooden plank for a slope, a protractor (angle measurer) and a stop-clock or timer.

Hypothesis

When toy cars are released at the top of an inclined plane (slope), the constant force of gravity pulls them down at the same speed every time.

Prediction

If a toy car is allowed to roll down an incline several times, it will always take the same amount of time to reach the bottom.

Investigation and recording data

1 Set up the plank with one end supported by the clamp on the stand and the other on the benchtop.

2 Adjust the plank so that it makes an angle of 30° with the benchtop.

3 Hold the toy car at the top of the slope with one hand and the stop-watch in the other.

4 Release the car and start the stop-watch.

5 Stop the stop-watch when the car reaches the bottom of the slope and record the time.

6 Repeat steps 3–5 four more times.

Examine the results

Examine the five records of the time taken for the car to travel down the slope. Calculate the speed in metres per second each time.

Conclusion

Compare the examination of your data with your hypothesis and prediction and draw a conclusion.

Is your conclusion limited in some way? Explain your answer.

What improvements could be made? Explain the changes that you suggest.

5 Scientists often calculate an average (mean) from their data where there is a range of results. This provides some reliability to the results. If your data shows a range of results, calculate the average (mean).

6 What would an anomalous result look like in the data?

7 If there is an anomalous result in the data, identify it and suggest how it might have happened.

One scientific enquiry often leads scientists to think of other questions which need further enquiries. For example, at the end of the previous enquiry, one question which could be in need of an answer is, 'Does the angle of the slope affect the speed of the toy car?' Try the following enquiry to find out.

Does the angle of slope affect the speed of the toy car?

You will need:

a clamp and stand, a toy car, a 1m long wooden plank for a slope, a protractor (angle measurer) and a stop-clock.

Hypothesis

Objects moving down a slope move more slowly than objects falling straight down, but if the slope is steep (almost straight down) objects should move faster.

Prediction

The steeper the slope, the faster an object travels down it.

Plan, investigation and recording data

1 Use the last enquiry to work out a plan to test the hypothesis and prediction.
2 If your teacher approves your plan, try it.
3 Record all the car runs down the slope, even those that may veer off to one side, or even those that fall off the slope before reaching the bottom.

Examining the results

Examine the data you have collected and look for a pattern. Identify any anomalous results and try to explain their cause.

Conclusion

Compare your evaluation with the hypothesis and prediction and draw a conclusion.

Is your conclusion limited in some way? Explain your answer.

What improvements could be made? Explain the changes that you suggest.

A question which might be asked after the last enquiry could be, 'Does the surface over which something travels affect its speed?'

Does the surface over which something travels affect its speed?

You will need:

the equipment from the previous enquiry, a selection of materials to put on the slope, for example sandpaper, a woollen cloth, shiny paper.

Hypothesis

The surface over which something travels will affect its speed.

Prediction

Base this on the materials you select for your experiment.

Plan, investigation and recording data

Select materials for the surfaces then make a plan of how you will use them in the experiment and make a table in which to record your data.

Examining the results

Examine your data and look for trends, patterns and anomalous results and note them down.

Conclusion

Use your findings from examining the data to draw a conclusion.

Is your conclusion limited in some way? Explain your answer.

What improvements could be made? Explain the changes that you suggest.

LET'S TALK

What other simple investigations could you carry out about the speed of objects (living or non-living)? How would you find the answers? Make a plan for one of them and, if your teacher approves, try it and report your findings to the whole class.

Ways to measure speed

In your investigations, you may have used a stop-clock or stop-watch to measure the time it takes an object to move a measured distance. While the stop-clock or stop-watch will work precisely every time, people do not. You may be a little slow in pressing the button at the start of the measurement and a little quick to press the button at the end. You may have noted these errors when drawing your conclusions. A more accurate way to make measurements is to replace you and your stop-clock with light gates.

Light gates

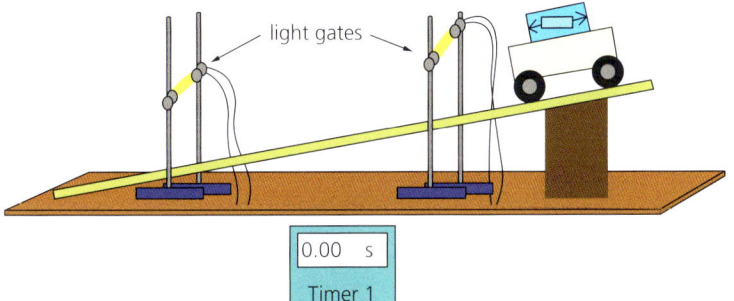

light gates

0.00 s

Timer 1

◀ **Figure 12.5** A set of light gates.

▲ **Figure 12.6** A speed-trap gun.

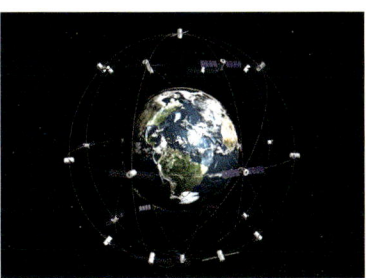

▲ **Figure 12.7** Satellites in orbit around the Earth.

In a light gate, a **light beam** shines onto a light-sensitive switch. When the beam of the light gate used at the start of a speed test is broken by an object passing through it, the switch starts an electronic stop-watch. When the beam of the light gate used at the finish of the speed test is broken by the object passing through it, it stops the stop-watch. The speed of the object is then found by dividing the distance between the light gates by the time, as measured by the electronic stop-watch.

The speed-trap gun

The speed-trap gun is a radar gun. (Radar stands for '**R**adio **d**etection **a**nd **r**anging' and describes a way of finding the position of objects and their speed using radio waves.) When the gun is fired at an approaching vehicle, a beam of radio waves travels to it through the air. Radio waves are waves with electrical and magnetic properties that transfer energy. These are reflected off the front of the vehicle and returns to a receiver on the gun. A computer in the gun compares the time difference between sending the beam and receiving it back from the vehicle and calculates the vehicle's speed.

Measuring speed from space

If a vehicle is equipped with an electronic device called a tracker, its position can be detected by a satellite in space and sent to a receiver back on Earth. When the vehicle moves from one position on the planet to another, the time taken to cover this distance is measured by a very accurate atomic clock (a time-keeper which uses the vibration of atoms to measure time) on the satellite.

8 Two people timed an object with a stop-watch. They each got a slightly different result. How could this be?

9 Which is more reliable: using a manual stop-watch or using light gates? Explain your answer.

Distance/time graphs

The distance travelled by an object over a period of time can be plotted on a graph called a **distance/time graph**. The distance covered by the object is recorded on the vertical axis and the time taken for the object to cover the distance is recorded on the horizontal axis.

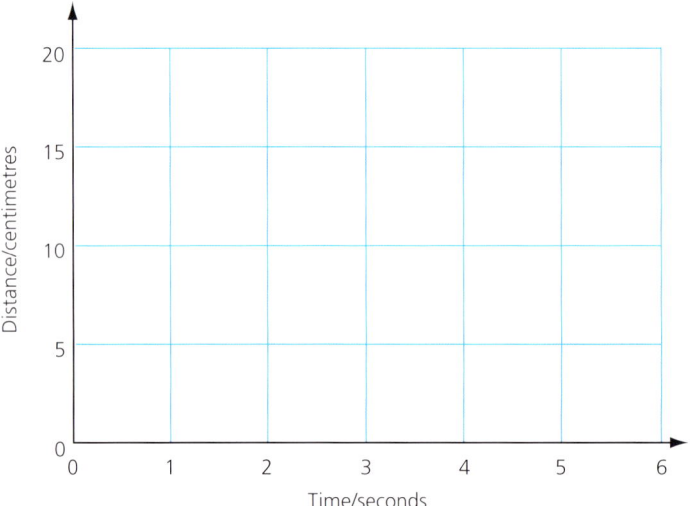

▲ Figure 12.8

If an object is not moving (it has no speed) the distance/time graph will look like Figure 12.9. This graph shows that the position of the object has not changed during the time period.

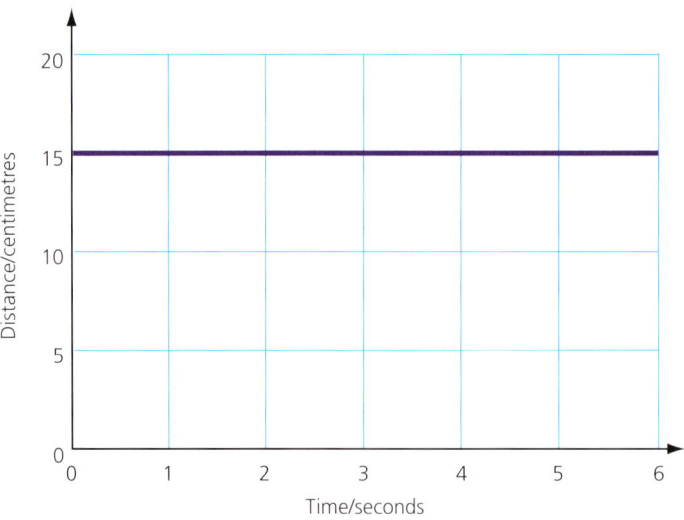

▲ Figure 12.9

If an object is moving at a constant speed, the distance/time graph will look like Figure 12.10. This graph shows that the object has moved at a constant speed of 5 cm every second.

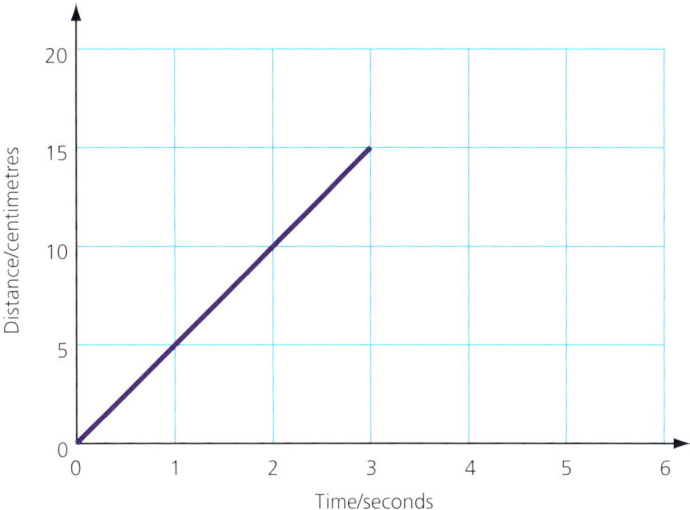

▲ **Figure 12.10**

Figure 12.11 shows the distance/time graph for an object which moved at a steady speed for two seconds then stopped and remained stationary for two seconds. If the object had been travelling faster in the first two seconds, the sloping line would be steeper. If the object had been travelling more slowly in the first two seconds, the sloping line would be less steep.

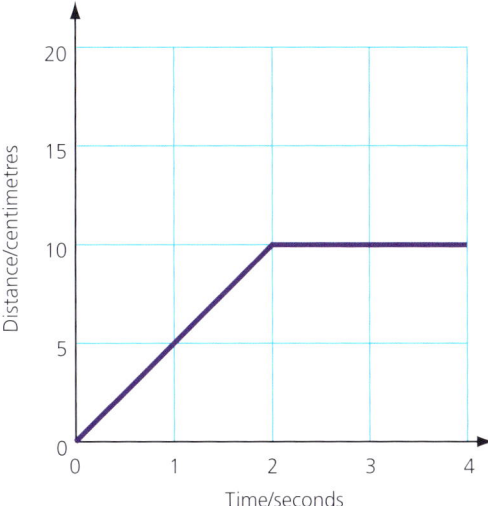

▲ **Figure 12.11**

The speed of an object, that is how far the object moves in a certain time, can be calculated from the distance time graph. For example, during the time it was moving the object in Figure 12.11 moved at 5 cm/s.

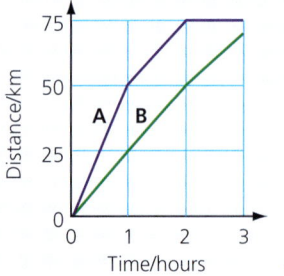

Time/hours

◀ **Figure 12.12**

10 Figure 12.12 shows the distance/time graph for two trucks, A and B, on an expedition across the Mongolian desert.
a i   How far did truck A move in the first hour of its journey?
   ii What was its speed during this time?
b How did the speed of truck A change in the second hour of its journey?
c Was truck B moving faster or slower than truck A in the first hour of its journey?
d What do you think might have happened to truck A in the third hour of the journey?

DID YOU KNOW?

The velocity is a description of the speed and direction of an object. For example, an aeroplane moving across the sky may have a velocity of 900 km/h in a direction due north.

Summary

✔ Speed = distance ÷ time
✔ Science in context: Some ways in which we can measure speed are: using a stop-clock (or timer), light gates, a light beam or a speed trap gun.
✔ We can plot the distance travelled by an object over time using a simple distance/time graph. These can show us if the object is travelling at a constant speed, slowing down or speeding up.

End of chapter questions

1 Arrange the words 'time distance speed' into the speed equation.
2 You can see a beetle about to run from under a chair across the room to a table three metres away. How could you measure the beetle's speed over the three metres?
3 Name two pieces of electrical equipment that can be used to measure speed.

4 A group of students investigated the movement of a model car. They set up a ramp at a height of 6 cm and let the car roll down it and across the floor (see Figure 12.13).

They measured the distance travelled by the car across the floor after it left the ramp. The experiment was repeated three more times with the ramp set at a height of 6 cm; then the ramp was reset at a different height and more of the car's movements were recorded. Table 12.2 shows the results of the investigation.

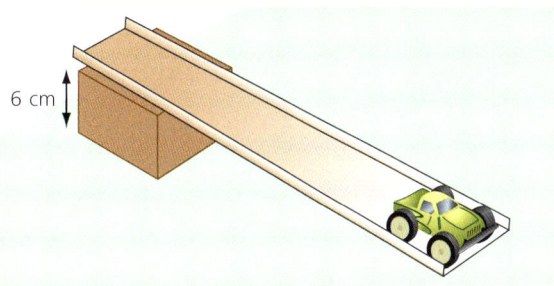

6 cm

◄ **Figure 12.13**

▼ **Table 12.2**

Height/cm	Distance/cm			
6	20	21	20	19
7	24	25	22	21
8	32	32	33	33
9	40	40	39.5	38
10	45	42	45	44
11	55	53	55	55
12	60	60	58	59
13	67	62	63	64

a How many times was the height of the ramp changed?

b How is an average (mean) calculated?

c Calculate the average (mean) distance travelled for each height of the ramp.

d Plot a graph to show the relationship between the height of the ramp and the distance travelled from the ramp by the car.

e What conclusions can you draw from your analysis of the results of this investigation?

5 Figure 12.14 shows the distance/time graph for a travelling object.

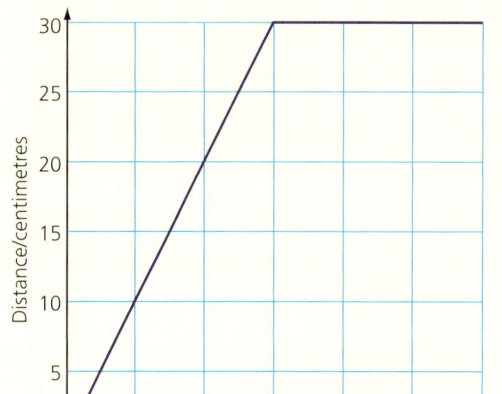

◀ **Figure 12.14**

a How far did the object move in the first 2 seconds?

b Why is the line horizontal between 3 and 6 seconds?

c Draw a distance/time graph for an object that travels at 15 cm/s for 1 second and then stops for 2 seconds.

d Draw a distance/time graph for an object that travels at 5 cm/s for 5 seconds and then stops for 1 second.

e Copy and complete this relationship: 'The steeper the line, the _____ the speed.'

6 A car travels 30 km in the first hour and then 40 km in the second hour. Draw a distance/time graph for the whole journey.

 Now you have completed Chapter 12, you may like to try the Chapter 12 online knowledge test if you are using the Boost eBook.

13 Forces

In this chapter you will learn:
- about Plimsoll lines (Science in context)
- about the effects of balanced and unbalanced forces on motion
- about exploring space (Science in context)
- about exploring the deep oceans (Science in context)
- to identify turning forces
- to calculate turning forces by knowing that moment = force x distance.

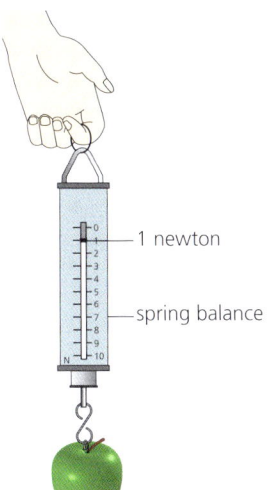

▲ **Figure 13.1** The apple pulls on the spring balance with a force equal to its weight.

Do you remember?

- What is the piece of equipment shown in Figure 13.1? What is the force pulling down on the apple and where is it pulling the apple to – the ground or the centre of the Earth?
- Look at Figure 13.2 and answer the following questions.
  - a What is the force pushing the kite in the air?
  - b What is the force pushing up on the boat?
  - c If one tug of war team drags the other across the ground, what force is acting between the feet of the losing team and the ground?
- An **applied force** is a force that is applied by something to something else. It can be a push or a pull. Use this information to describe the applied force when
  - a you move this book across your desk
  - b a car tows a trailer.

▲ **Figure 13.2** Forces act in many ways.

LET'S TALK

▲ **Figure 13.3** In cricket, applied forces make the ball move, stop, change direction, change speed and even change shape.

Watch a short video recording of part of a football, tennis or cricket match and identify the times when an applied force occurs and what it causes in the game.

135

Can your power to estimate forces be improved?

You will need:

a force meter (Newton meter), a plastic bag and a selection of small objects.

Hypothesis
The power to estimate accurately increases with experience.

Prediction
If a person tries a number of estimates over time, the accuracy of estimating will improve.

Investigation and recording data
1   Attach a plastic bag to the hook of the force meter (Newton meter).
2   Hold one of the small objects and estimate the force of gravity pulling it down. Record your estimate.
3   Put the object in the bag and read the size of the force on the scale and record it.
4   Repeat steps 2 and 3 with each object.

Examining the results
Look at your estimates and accurate measurements. Can you see a pattern in the results that supports the prediction? Were there any anomalous results? If so, identify them and offer an explanation for each one.

Conclusion
Compare the examination of your data with the hypothesis and prediction and draw a conclusion.

Is your conclusion limited in some way? Explain your answer.

What improvements could be made? Explain the changes that you suggest.

LET'S TALK

This enquiry is designed to look for a pattern in the results, but what may prevent a prediction or an estimate from being accurate? Look for anomalous results in the data and if you find any, explain their cause.

We know there are many kinds of forces, and there is more than one force acting on everything around you.

Balanced forces

If the forces are balanced, an object does not move, or it moves at a constant speed. An object that does not move is said to be a **stationary** object.

1 In a force diagram, what two things does an arrow tell you about the force?

The balanced forces on a stationary object
Weight and normal force
When you stand still, you do not rise above the ground or sink into it, because of two balanced forces acting on you. Your **weight** acts downwards on the ground because of gravity, and the ground exerts a normal or supporting force upwards on the soles of your shoes, as shown in Figure 13.4a. The strength of the force acting downwards is the same as the strength of the force acting upwards. The forces balance each other. This can be represented more clearly with a force diagram, as shown in Figure 13.4b.

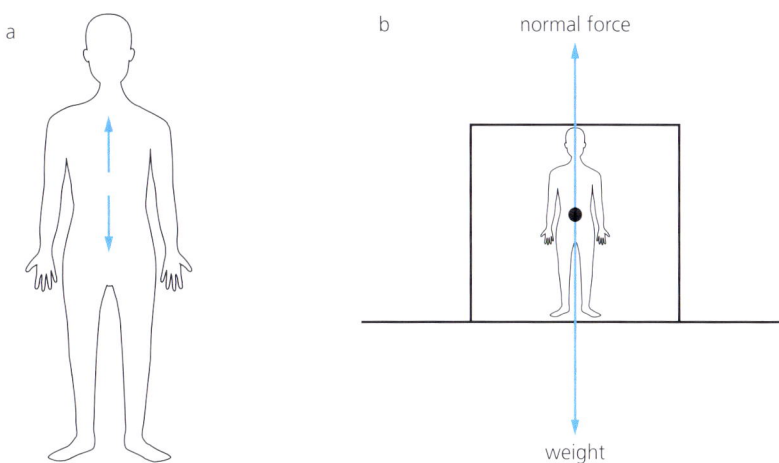

▲ **Figure 13.4a** Gravity and a supporting force acting in balance; **b** A force diagram can show this more clearly.

CHALLENGE YOURSELF
Look around you and choose five objects which are not moving. Draw force diagrams for each one.

Weight and upthrust
On the still water, a stationary floating object is held in position by the balanced forces of weight and upthrust. The buoy in Figure 13.5a is carrying an automatic weather station. It floats because its weight downwards on the water is balanced by the upthrust of the water. This can be represented in a force diagram, as Figure 13.5b shows.

▲ **Figure 13.5a** A floating buoy with an automatic weather station; **b** its force diagram.

Plimsoll lines

In the nineteenth century, trade between countries increased and many kinds of goods were transported by sea. Some ships were loaded with so much cargo that they could barely float, and when they encountered stormy conditions, they sank. To reduce the number of shipping disasters and to save the lives of sailors, the politician Samuel Plimsoll (1824–1898) brought in a law in 1876 requiring each ship to be marked with a line showing the maximum level to which the water should rise up the sides in dock. This prevented any ship from being overloaded and reduced the chance of it sinking in a storm. These lines became known as Plimsoll lines.

▲ **Figure 13.6** Plimsoll lines on the side of a boat.

The symbol on the right in Figure 13.6 shows the loading limit in different situations. It shows how the level of the water will change as the ship moves into different kinds of water. These changes in level are due to the different densities of the different kinds of water, which leads to differences in upthrust. For example, the upthrust on a fully loaded ship in cold, dense sea-water in the North Atlantic Ocean in winter is so strong that a ship only sinks to the level marked WNA. If the fully loaded ship sailed into a warm, **tropical**, freshwater river it would sink to the level marked TF.

2 What effect does the temperature of the water have on the upthrust of the water on a ship?

3 How does the presence of salt in water affect the upthrust of the water on a ship?

4 What might happen to a ship that was loaded to the line in cold salt water (W) and then sailed into warm fresh water? Explain your answer.

The balanced forces on an object moving at constant speed

Imagine that you are in a go-kart, like the one shown in Figure 13.7a, but it is not moving. Your weight and the weight of the go-kart are pushing down on the ground, and the normal force exerted by the ground is pushing back up to support you and the kart. The forces are balanced. This can be represented in a force diagram, as Figure 13.7b shows.

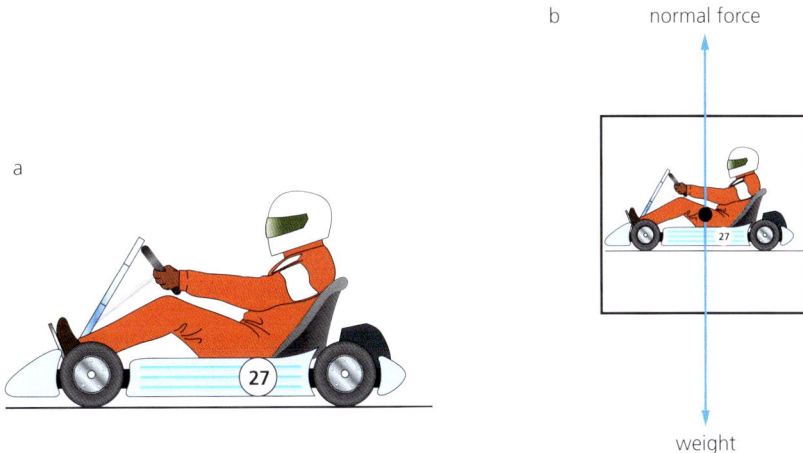

▲ **Figure 13.7a** A stationary go-kart; **b** its force diagram.

If you start your go-kart and get it moving at a constant speed, there are two more forces acting on you. They are air resistance and the driving force from the engine (see Figure 13.8a). This can be represented in a force diagram, as Figure 13.8b shows.

CHALLENGE YOURSELF

A fly moves forwards by the driving force of its flight muscles moving its wings. Imagine a fly moving across the room in front of you at a constant speed. Draw a forces diagram to describe its movement.

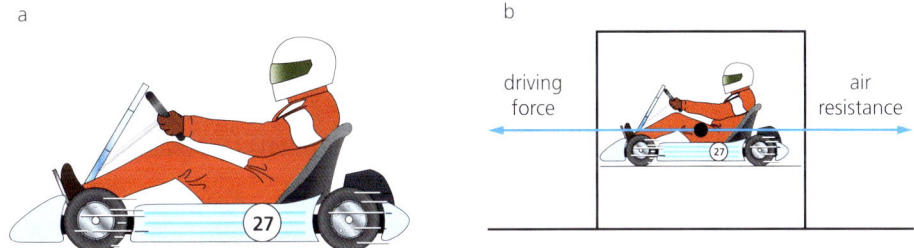

▲ **Figure 13.8a** A go-kart moving at constant speed; **b** its force diagram.

Science in context

Exploring space

One of the ways that we explore space is to send out space probes across the solar system, and even into deep space. The space probes do not need rocket engines to be permanently working in order to keep moving. The reason for this is due to the work of Isaac Newton (1642–1727), who studied forces and summarised his work in **three laws of motion**. The first law states that everything stays at rest or moves steadily in a straight line, unless another force pushes or pulls on it.

When a space probe is launched, it leaves the Earth's atmosphere and its rocket engines push it out into space. When the engines are switched off, it keeps moving, because in space there is no air resistance to push on it and slow it down. As no other force pushes or pulls on the spacecraft, it could keep travelling across the universe forever.

The Voyager 1 spacecraft was launched in 1977 and it is still travelling through space at 61 000 km/hr. In 40 000 years, it will pass a star called AC+79 3888, which is about 17.5 light-years away.

5 A light-year is the distance that light can travel in a year: 9 500 000 000 000 kilometers. How far is star AC+79 3888 away in kilometers?

6 If a spacecraft comes close to a planet or a star, will it continue to move steadily in a straight line? Explain your answer.

▲ **Figure 13.9** Voyager 1.

Unbalanced forces

In the last section, you were asked to imagine being in the go-kart when it was rest and at constant speed. In both those situations, the forces acting on you and the kart were balanced. You may wonder at how the change from not moving to moving at a constant speed was achieved. It was achieved by unbalanced forces in the following way.

When you press a pedal on the kart, called the accelerator pedal, the engine pushes you and the kart forwards. The drive force produced by the engine is more than the air resistance. This means the forces are unbalanced and the drive force moves you forwards (see Figure 13.10a). This can be represented in a force diagram as Figure 13.10b shows.

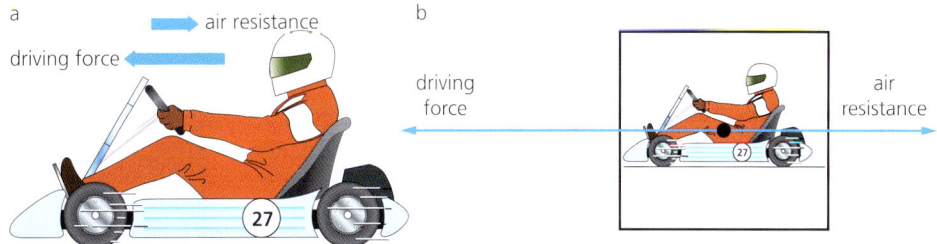

▲ **Figure 13.10a** The drive force is greater than the air resistance, so the kart moves forwards; **b** the force diagram showing this.

After travelling in your go-kart at constant speed for a while, you may decide to stop and take your foot off the accelerator pedal. When you do this, the size of the drive force of the engine becomes smaller than the air resistance, and you slow down.

There is a second pedal on the kart, called the brake. If you press it as you slow down, a **friction** force is generated by a pad which presses against the wheels. This force pushes in the same direction as the air resistance, and together the two forces bring you and your go-kart to a halt.

▲ **Figure 13.11** Real-life go-karting.

Sinking

When an object floats on water, its weight is balanced by the upthrust. If something happens to increase the weight of the object, so much that it can no longer be balanced by the upthrust, the object sinks. A boat floats because its weight is balanced by the upthrust, but if there is a hole in the boat, water enters and increases the weight of the boat so much that the upthrust can no longer keep it afloat, and the boat sinks.

Some objects have a weight that is far greater than the upthrust pushing on them, and they sink as soon as they reach the surface of the water.

7 Draw a force diagram of a boat that has sunk below the surface of the water.

Exploring the deep oceans

A bathyscaphe is used to explore the deep oceans. It has air tanks which help it float at the surface so that its weight is less than the upthrust. When the crew are ready, they let the air tanks fill with water, which increases the bathyscaphe's weight so much that the upthrust can no longer balance it, and the bathyscaphe sinks. There are engines on board that can move the crew around as they explore the ocean bed. When the survey is complete, they slowly release iron pellets, which reduces the weight of the bathyscaphe, and it rises gradually back to the surface.

▲ **Figure 13.12** The crew make their observations from inside the white sphere.

Observing balanced and unbalanced forces

Sometimes a simple observation can lead to a scientific explanation if some scientific knowledge is known. Undertake this scientific enquiry to generate some observations which are explained in terms of balanced and unbalanced forces in the text that follows.

What can you observe about upthrust?

You will need:

an empty plastic bottle with the top screwed on and a deep bowl of water.

Investigation and observation

1 Place the bottle on the water's surface and observe.
2 Gently push down on the bottle and observe what happens with your senses of sight and touch.
3 Quickly move your hand away from the bottle and observe what happens.
4 Record your observations.

Examining the results

Look at the results and describe your observations.

Conclusion

Use the information in the three points that follow this box (on the next page) to help you draw a conclusion.

Your observations can be explained in the following way.

- The bottle resting on the water's surface has upthurst pushing it up and weight pushing it down. The two forces are balanced and the bottle stays in position.
- When you push down on the bottle you are increasing the downward force and this is balanced with a strengthening of the upthrust which you feel as you push down.
- When you release the bottle, the downward force is reduced to just the bottle's weight, but the upthrust is much stronger. The forces are unbalanced and the bottle rises to the surface and comes to rest when the forces on it are balanced again.

Turning forces

The turning effect of forces

A force can be used to turn an object in a circular path. For example, when you push down on a bicycle pedal, the cog-wheel attached to the crank-shaft (the metal bar that connects the cog wheel and the pedal) turns round. A cog wheel (also known as a gear wheel) is a wheel which has teeth that can be used to move other gear wheels and chains on bicycles.

A nut holding the hub of a bicycle wheel to the frame is turned by attaching a spanner to it and exerting a force on the other end of the spanner in the direction shown in Figure 13.13.

A device that changes the direction in which a force acts is called a **lever**. It is composed of two arms and a **fulcrum** or **pivot**. The lever also acts as a force multiplier. This means that a small force applied to one arm of the lever can cause a large force to be exerted by the other arm of the lever.

For example, a crowbar is a simple lever that is used to raise heavy objects. One end of the crowbar is put under a heavy object and the crowbar is rested on the fulcrum (Figure 13.14). When a downward force is applied to the long arm of the crowbar, an upward force is exerted on the heavy object. A small force acting downwards at a large distance from the fulcrum on one side produces a large force acting upwards a short distance from the fulcrum on the other side.

The force applied to the lever to do the work is called the **effort**. It opposes the force that is resisting the movement, called the **load** (Figure 13.14).

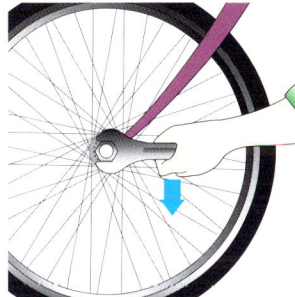

▲ **Figure 13.13** Tightening a nut.

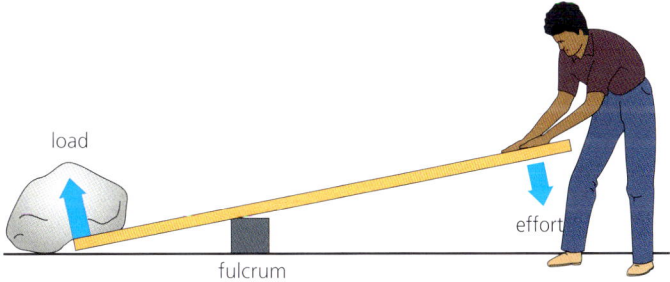

load

effort

fulcrum

▶ **Figure 13.14** Using a lever: a simple crowbar.

8 Where will the strongest force be exerted by scissor blades to cut through a piece of material? Explain your answer.

9 Why can a lever be described as a force multiplier?

A pair of pliers or a pair of scissors (Figure 13.15) are made from two levers. When pliers are used to grip something, a small force applied to the long handles produces a large force at the short jaws. In a pair of scissors, the arms on one side of the pivot have a sharp edge. When a force is applied to the handles on the other arm, a force is generated on the sharp edges which can cut through paper and cloth.

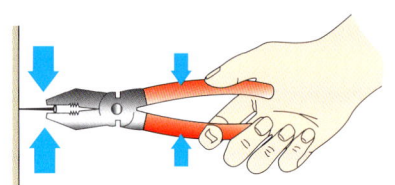 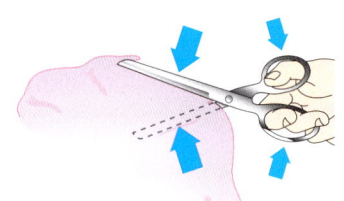

▲ **Figure 13.15** Using pliers and scissors.

▲ **Figure 13.16** A see-saw.

Moments

The turning effect produced by a force around a fulcrum is called the **moment** of the force. This is best understood by considering a see-saw, as shown in Figure 13.16.

The direction of the moment is usually specified as clockwise (the load in Figure 13.14) or anti-clockwise (the effort in Figure 13.14) about the fulcrum. The size of the moment is found by multiplying the size of the force by the distance between the point at which the force acts and the fulcrum (in Figure 13.16, these are the distances from the ends to the fulcrum).

The moment of a force can be shown as an equation:

moment of force = force × distance from the fulcrum

The moment is measured in newton metres (Nm). The moment of the force applied to one arm of a lever is equal to the moment of the force exerted by the other arm. For example, a 100N force applied downwards 2m from the fulcrum produces a 200N force upwards 1m from the fulcrum on the other arm.

In the case of the see-saw, the moment of the weight on one arm must equal the moment of the weight on the other arm for the see-saw to balance. From this observation, Archimedes (287–212BC), living in Greece, constructed his **law of moments**, which is sometimes called the **law of the lever** or the **principle of moments**. This law or principle states that when a body is in equilibrium (or balance), the sum of the clockwise moments about any point (such as the fulcrum) equals the sum of the anti-clockwise moments about that point.

10 What will happen if the anti-clockwise moment of a see-saw is greater than the clockwise moment?

If the sum of the clockwise moment is greater or less than the sum of the anti-clockwise moment, the see-saw is not balanced. If the clockwise moment is greater than the anti-clockwise moment, it will pull that side of the see-saw down and the other side of the see-saw will rise.

Calculating and recording moments

Figure 13.17 shows the equipment to use when investigating moments.

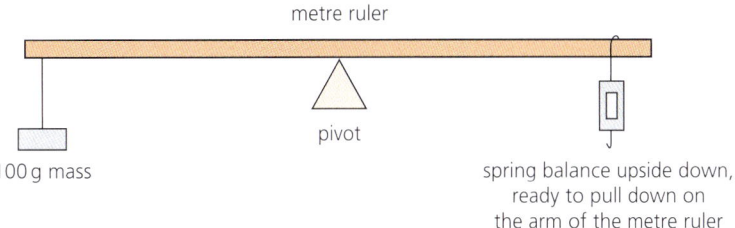

metre ruler

pivot

100 g mass

spring balance upside down, ready to pull down on the arm of the metre ruler

▲ **Figure 13.17**

In an investigation, a mass is placed on one end of the metre ruler (metre stick) and a spring balance is attached, upside down, to the other end to pull it down to make the metre ruler (metre stick) horizontal.

When a mass is hung on the metre ruler (metre stick), its distance from the pivot is recorded. The value of the force of the mass is multiplied by the distance from the pivot to calculate the moment of the force. For example, a 100 g mass exerts a downward force of 1 N and when this is multiplied by 20 the moment is calculated as $1 \times 20 = 20$ Nm and is recorded in a table like Table 13.1.

The spring balance can then be placed at 10 cm from the pivot on the other end and can be pulled down until the metre ruler (metre stick) is horizontal and the force exerted by the spring balance (2 N) is recorded and the moment is calculated as $2 \times 10 = 20$ Nm.

Set up	Moment of force on left end (Nm)	Moment of force on right end (Nm)
1	20	20

▲ **Table 13.1**

Use this information to help you investigate moments in the following enquiry.

Can you investigate moments?

Work safely

Take care not to drop the weights on your feet.

You will need:

a metre ruler (metre stick), a 100 g mass on a thread, a spring balance and thread, and a suitable pivot such as a triangular wooden block, a partner to help in holding the equipment as you set it up.

Hypothesis

Changing the size of the force and its position along the long arm of a beam can keep the beam horizontal.

Prediction

As the force is increased, its distance from the fulcrum to keep the beam horizontal also increases.

Investigation and recording data

1. Make a table like the one in Table 13.1 but with more rows in which to record your data.
2. Figure 13.17 shows how the equipment is set up for this enquiry. Set up the pivot and rule by placing the pivot on the corner of a bench or table so that both arms of the metre ruler (metre stick) have open space below them.
3. Carefully hang the mass on the left arm of the metre rule (metre stick) while the rule is being held horizontally.
4. Calculate the moment of the force in the left arm of the ruler (stick) and enter it in your table.
5. Calculate the force needed at a point 50 cm from the pivot on the long arm to balance the mass.
6. Check your answer by fixing a spring balance 50 cm from the pivot on the long arm and pulling up until the metre rule is horizontal. Measure the size of the force and compare it with your calculation.
7. Calculate the force needed at other distances along the long arm to keep the metre rule horizontal, and check with the spring balance.
8. Note down and present your results in a table.
9. Plot a graph of your readings of measured force versus distance from the pivot.

Examining the results

What does the graph show? Write a sentence.

---

11. What is the moment of a 100 N force acting on a crowbar:
    a 2 m from the fulcrum
    b 3 m from the fulcrum
    c 0.5 m from the fulcrum?

12. A 100 N force acting on a lever 2 m from the fulcrum balances an object 0.5 m from the fulcrum on the other arm.
    a What is the weight of the object in newtons?
    b What is its mass in kg?

Conclusion

Compare your examination of the graph with the hypothesis and prediction and draw a conclusion.

Is your conclusion limited in some way? Explain your answer.

What improvements could be made? Explain the changes that you suggest.

Summary

✔ Science in context: Plimsoll lines were invented in the nineteenth century to ensure ships were not overloaded and unsafe in storms.
✔ If forces are balanced, an object does not move, or it moves at a constant speed.
✔ If forces are unbalanced, an object will speed up or slow down.
✔ A force can be used to turn an object in a circular path.
✔ The turning effect produced by a force around a fulcrum is called the moment of the force.
✔ Moment of force = force × distance from the fulcrum.

End of chapter questions

1 When you sit on a chair, what force acts
   a downwards
   b upwards?
2 What information does a force arrow give you?
3 Imagine you are sitting in a boat on a lake.
   a i   What force do you exert on the water?
      ii What force does the water exert on your boat?
   b Why is your boat not sinking?
   c You hit a rock and stop. You see that there is a hole in your boat and the water is coming in. Explain, in terms of forces, what will happen to your boat if you do not block the hole.
4 You are in a spacecraft moving fast and the engines switch off. What happens next? Explain your answer in terms of forces.

▲ **Figure 13.18**

5   In Figure 13.18, identify places where there are
    a   balanced forces
    b   unbalanced forces
    acting on a person or object.
6   Why does a stone sink when it falls in a river?
7   What is another name for 'pivot'?
8   What is the difference between the load and the effort?
9   If you sit on one end of a see-saw, how can you calculate the moment of your force?
10 Is it possible to balance a mass of weight 5 N with a mass of weight 15 N on a model see-saw with 10 cm arms? Explain your answer.
    Suggest where you might place each mass to get an exact balance.

CHALLENGE YOURSELF

Is the upthrust generated in water the same as in cold cooking oil? How could you use a straw with a modelling clay weight to find out?

Now you have completed Chapter 13, you may like to try the Chapter 13 online knowledge test if you are using the Boost eBook.

14 Pressure and diffusion

In this chapter you will learn:
- that pressure = force ÷ area
- how pressure is caused by the action of a force from a substance on an area
- how to explain pressures in gases and liquids by referring to the arrangement of their particles
- about the thought experiment (Science in context)
- about reservoirs and dams (Science in context)
- about hydraulic equipment (Science in context)
- about ear popping (Science extra)
- about hovercraft (Science in context)
- how the diffusion of gases and liquids is due to the movement of particles.

Do you remember?

- What is the particle model?
- How are the particles arranged in a model of a
  - a solid
  - b liquid
  - c gas?
- Describe the forces acting on a block of wood when it
  - a rests on the ground
  - b is pushed along the ground
  - c floats on water
  - d moves through the air after being thrown.
- What is the unit for measuring mass?
- What is the unit for measuring weight?
- What is the difference between mass and weight?
- What would you put on a force diagram to show the forces acting on a wooden block on the ground?

A phenomenon is an event that can be observed by the senses. A rainbow, for example, is a phenomenon that you observe with your eyes. Scientists study phenomena to find out about them – what causes them and what they can do. Sometimes, when studying different phenomena, they find that a previous piece of knowledge can be used to explain them.

Pressure and diffusion in liquids and gases are two examples of phenomena that can use the same piece of previous knowledge to explain them. First we will look at pressure.

If you hold out your left hand with the palm upwards and press down on your fingertips with the fingertips of your right hand, what do you feel? You may say that there is a force pushing down from your right hand or you may say that your fingertips are feeling a pressure. Both explanations would be correct. Now repeat with the left hand pushing on the right hand. Is there a difference? You may answer using words such as 'force' and 'pressure', but what is the connection between them? You will find out in this chapter.

Pressure on a surface

When a force (such as the push of your hand) is exerted over an area (such as the area of your fingertips) we describe the effect in terms of pressure. Pressure can be defined by the equation:

$$\text{pressure} = \frac{\text{force}}{\text{area}}$$

The SI unit for pressure is N/m² (newtons per square metre) but it can also be measured in N/cm² (newtons per square centimetre).

An object resting on a surface exerts pressure on the surface because of the object's weight. **Weight** is the force produced by gravity acting on a solid, a liquid or a gas, pulling the material downwards towards the centre of the Earth. The weight acts on the mass of that material. For example, the weight of a solid cube acts on that cube (Figure 14.1).

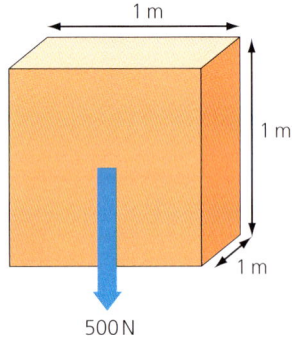

1 m
1 m
1 m
500 N

▲ **Figure 14.1** Weight acting on a cube of material.

The cube pushes down on the ground (or other surface that it rests on) with a force equal to its weight. The pressure that the cube exerts on the ground is found by using the equation above. For example, if the cube has a weight of 500 N and the area of its side is 1 m², the pressure it exerts on the ground is:

$$\text{pressure} = \frac{500}{1}$$

$$= 500 \, \text{N/m}^2$$

If the cube has a weight of 500 N and the area of its side is 2 m², the pressure it exerts on the ground is:

$$\text{pressure} = \frac{500}{2}$$

$$250 \, \text{N/m}^2$$

1 What is the pressure exerted on the ground by a cube which has a weight of 600 N and a side area of:
   a 1m²   b 3 m²?
2 What is the pressure exerted on the ground by an object that has a weight of 50 N and a surface area in contact with the ground of:
   a 1 cm²   b 10 cm²   c 25 cm²?
3 a What pressure does a block of weight 600 N and dimensions 1m × 1m × 3 m exert when it is:
      i   laid on its side   ii stood on one end?
   b Why does it exert different pressures in different positions?

An object exerts a pressure on the ground according to the area of its surface that is in contact with the ground. For example, a block with dimensions 1 m × 1 m × 2 m and a weight of 200 N will exert a pressure of 200 ÷ 1 = 200 N/m² when it is stood on one end (Figure 14.2a) but a pressure of only 200 ÷ 2 = 100 N/m² when laid on its side (Figure 14.2b).

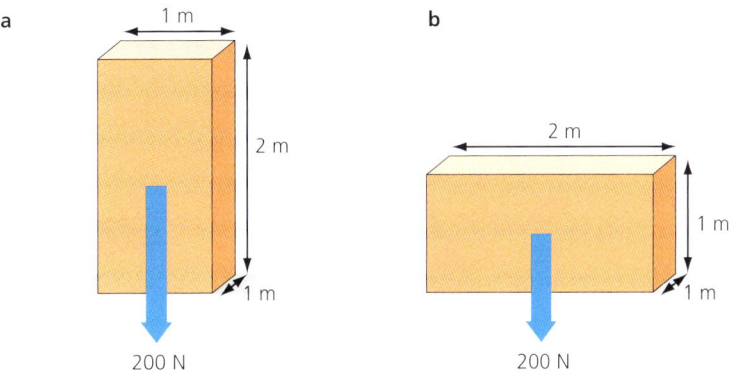

▲ **Figure 14.2** The weight acting on a block in two positions.

Your weight acting downwards causes you to exert a force on the ground through the soles of your shoes. If you lie down, this force acts over all the areas of your body in contact with the ground. These areas together are larger than the areas of the soles of your shoes and you therefore push on the ground with less pressure when lying down than when you are standing up.

▲ **Figure 14.3** The force you exert downwards acts over a larger area when you lie down.

4   When going out on the snow, drivers in Iceland let their tyres down until they are very soft. The tyres spread out over the surface of the snow as they drive along. Why do you think the drivers do this?

Reducing pressure

When people wear skis, the force due to their weight acts over a much larger area than the soles of a pair of shoes. This reduces the pressure on the soft surface of the snow and allows the skier to slide over it without sinking.

▲ **Figure 14.4** Skis stop you from sinking into the snow.

Increasing pressure

Studs

Boots for sports like soccer, rugby and hockey have studs on their soles. They reduce the area of contact between your feet and the ground. When you wear a pair of these boots, your downward force acts over a smaller area than the soles of your feet and you press on the ground with increased pressure. Your feet sink into the turf on the pitch and grip the surface more firmly. This makes it easier to run without slipping while you play the game.

▲ **Figure 14.5** The spikes stop the hurdler from slipping on the track. In a similar way, the studs on a soccer boot help the player to grip the turf.

Pins and spikes

A drawing pin (or thumb tack) is a small pin with a broad head that can be pushed into surfaces to hold up light materials, such as paper or cardboard. When you push a drawing pin into a board, the force of your thumb is spread out over the head of the pin so the low pressure does not hurt you. The same force, however, acts at the tiny area of the pin point. The high pressure at the pin point forces the pin into the board.

Hurdlers use sports shoes that have spikes in their soles. The spike tips have a very small area in contact with the ground. The weight of the hurdler produces a downward force through this small area and the high pressure pushes the spikes into the hard track, so the hurdler's feet do not slip when running fast.

5 A girl wearing trainers does not sink into the lawn as she walks across it but later, when she is wearing spiked shoes for sprinting, she sinks into the turf. Why does this happen?

How does the area of a block in contact with a surface affect the pressure?

You will need:

a bowl of sand, two blocks of wood (one with an end with a smaller surface area than the other), a mass to provide weight to push on the blocks, a camera and a ruler to measure the depth of the hole produced in the sand.

Hypothesis

The area of an object in contact with a surface affects the pressure on the surface.

Prediction

An object with a smaller area in contact with a surface will exert greater pressure on the surface and sink further into it.

Investigation and recording data

1 Smooth the surface of the sand in the bowl.
2 Place one end of the smaller block on the sand surface.
3 Place the mass on the block and observe the block sinking into the sand.
4 Carefully remove the mass and the block.
5 Measure the depth of the hole.
6 Repeat steps 1–5 with the larger block.
7 Make a note of your results by measuring the depth of each hole and photographing it.

Examining the results

Compare your measurements and the photographs of the holes made by the blocks.

Conclusion

Compare your examination of the results with the hypothesis and prediction.

CHALLENGE YOURSELF

How could you improve the experiment on surface area to make the data more reliable? What extra equipment would you use? Make a plan and, if your teacher approves, try it.

Scientists sometimes produce simple experiments to test a hypothesis like the enquiry above. They then refine the experiment to check the reliability of the data.

Knives

As we have seen, high pressure is made by having a large force act over a small area. The edge of a sharp knife blade has a very small area but the edge of a blunt knife blade is larger. If the same force is applied to each knife, the sharp blade will exert greater pressure on the material it is cutting than the blunt knife blade, and will therefore cut more easily than the blunt blade.

▲ **Figure 14.6** Knives cut well when they are sharp because of the small surface area of the blade.

Particles and pressure

Matter is made from particles. In solids, the particles are held in position. In liquids, the particles are free to move around each other. In gases, the particles are free to move away from each other (see Figure 14.7).

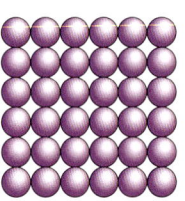

solid particles vibrate to and fro

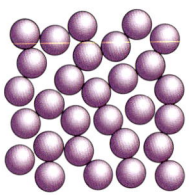

liquid particles have some freedom and can move over each other

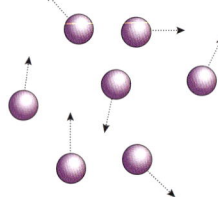

gas particles move freely and at high speed

▲ **Figure 14.7** Arrangement of particles in a solid, a liquid and a gas.

Pressure in liquids

In a solid object, the pressure of the particles acts through the area in contact with the ground. In a liquid, however, the pressure of the particles acts not only on the bottom of the container but on the sides too (Figure 14.8).

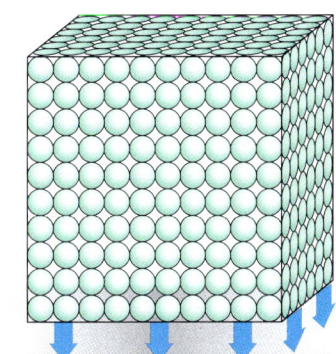

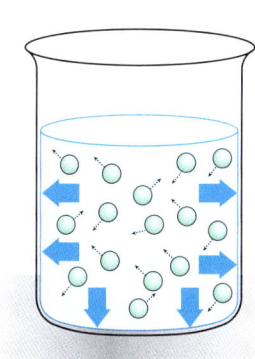

▲ **Figure 14.8** Pressure exerted by: **a** particles in a solid block; **b** particles in a liquid.

Pressure and depth in a liquid

Scientists sometimes begin answering a question by making a thought experiment before making a real experiment. Here is an example.

Science in context

The thought experiment

The question to answer is: 'Does the pressure in a liquid depend on its depth, and how can we find out?'

We can begin to answer this by imagining a layer of water in the bottom of a beaker. It is so thin that it is only one particle thick, like the row of particles at the bottom of the beaker in Figure 14.8. We can see that the particles push on the bottom of the beaker, and those next to the sides of the beaker also push on the sides.

Next we add row upon row of particles, until it looks like the whole beaker of particles in Figure 14.8. We can imagine that, as the layers of particles build up, they push with an even greater force on the bottom of the beaker, but we may be unsure if they also push with an even greater force on the sides of the beaker.

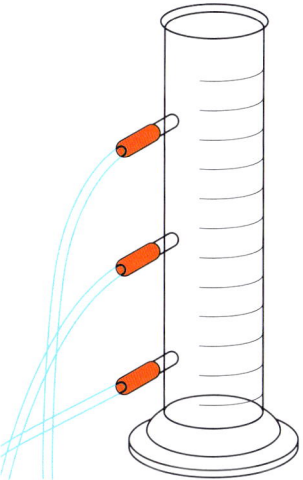

One way to test this thought would be to make the beaker very tall and have holes in its side. If the water is under a high pressure, it will shoot a long way through the hole, but if it is at a low pressure, it would not shoot out as far. The equipment to test this idea could be as shown in Figure 14.9. This piece of equipment is called a **spouting cylinder**. The rubber tubes could be nipped shut with clamps to prevent the water escaping while the cylinder is being filled up.

▲ **Figure 14.9** A spouting cylinder.

A thought experiment is often followed by a real experiment, like the one in the next enquiry.

Does depth of water affect its pressure?

You will need:

a spouting cylinder or a tall plastic bottle with three holes in it down one side (the holes are stopped with modelling clay which is then removed to release the jets of water), a ruler and a tray to collect the water.

Hypothesis

If a spouting cylinder, like the one in Figure 14.9, is filled with water and the clamps are released to let the water out, there should be a difference in the jets of water due to the difference in water pressure in the cylinder.

Prediction

Use the information in the thought experiment to make a prediction about how the jets of water may differ.

Investigation and recording data

1 Fill the spouting cylinder with water.
2 Stand the spouting cylinder in the tray at one side with the holes pointing towards the middle.
3 Open the top hole and observe the distance at which the water falls into the tray. Measure and record the distance.
4 Repeat step 3 with the two remaining holes.

Examining the results

Compare the data you have recorded.

Conclusion

Compare the examination of your data with the hypothesis and prediction and draw a conclusion.

Is your conclusion limited in some way? Explain your answer.

What improvements could be made? Explain the changes that you suggest.

6 How would the length of a jet compare with the others if it was to come from a spout
   a between the top and middle spouts
   b between the middle and bottom spouts
   of the spouting cylinder?

The relationship between pressure and depth has been investigated many times, and a diagram summarising the results has been drawn. It shows that the greatest pressure is at the bottom of the spouting cylinder, which produces the longest jet of water, while the pressure at the top is much less, which produces a shorter jet.

Science in context

Reservoirs and dams

Water is a vital resource and in many places it is stored in reservoirs for use in towns, cities, farms and for generating electricity.

A reservoir is made by placing a dam across a river. The water is stopped from flowing down the river and builds up into a large body of water – the reservoir, which is like a lake.

◀ **Figure 14.10** A reservoir made in a valley by a huge dam.

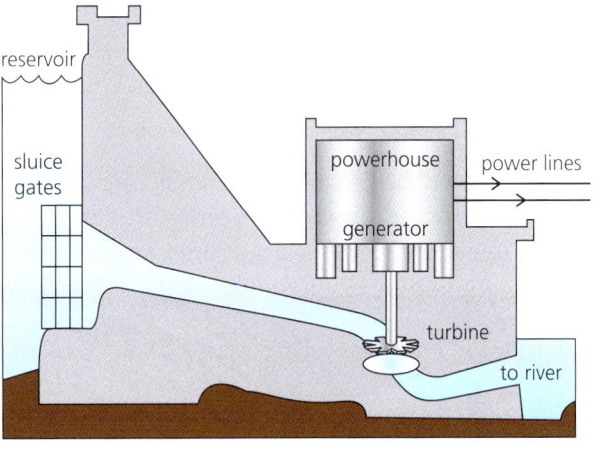

◀ **Figure 14.11** A cross-section of a hydro-electric power station.

7 Why is the dam thinner at the top and thicker at the bottom?

8 Why is the water drawn from the bottom of the dam rather than the top?

This hydro-electric power station uses the force of flowing water to spin a turbine, which is connected to an electrical generator. As the turbine spins, the electicity is produced in the generator and conducted away on power lines to where it is needed.

Hydraulic equipment

If pressure is applied to the surface of a liquid in a container, the liquid is not squashed, because the particles in a liquid are already touching each other. It transmits the pressure so that it pushes on all parts of the container with equal strength.

In hydraulic equipment, a liquid is used to transmit pressure from one place to another. In Figure 14.12, a small force pushes piston A a long way down its tube and generates a high pressure. This high pressure is transmitted to all parts inside the equipment. At piston B, the high pressure generates a larger force than at piston A, which is shown by the length of the arrows. This force pushes the larger area of piston B upwards, but by a much smaller distance than that travelled by piston A.

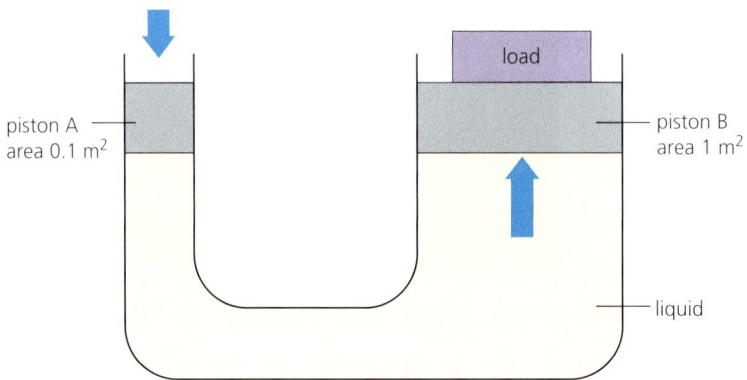

piston A
area 0.1 m^2

load

piston B
area 1 m^2

liquid

▲ **Figure 14.12** A simple hydraulic system.

A car may be raised with a small force by using a hydraulic jack. When a small force is applied to a small area of the liquid in the jack, a larger force is released across a larger area and acts to raise the car.

▲ **Figure 14.13** This car has been raised into the air for repairs by a hydraulic jack.

The brake system on a car is a hydraulic mechanism. The small force exerted by the driver's foot on the brake pedal is converted into a large force acting at the brake pads. This results in a large frictional force that makes it harder for the wheels to turn and so stops the car.

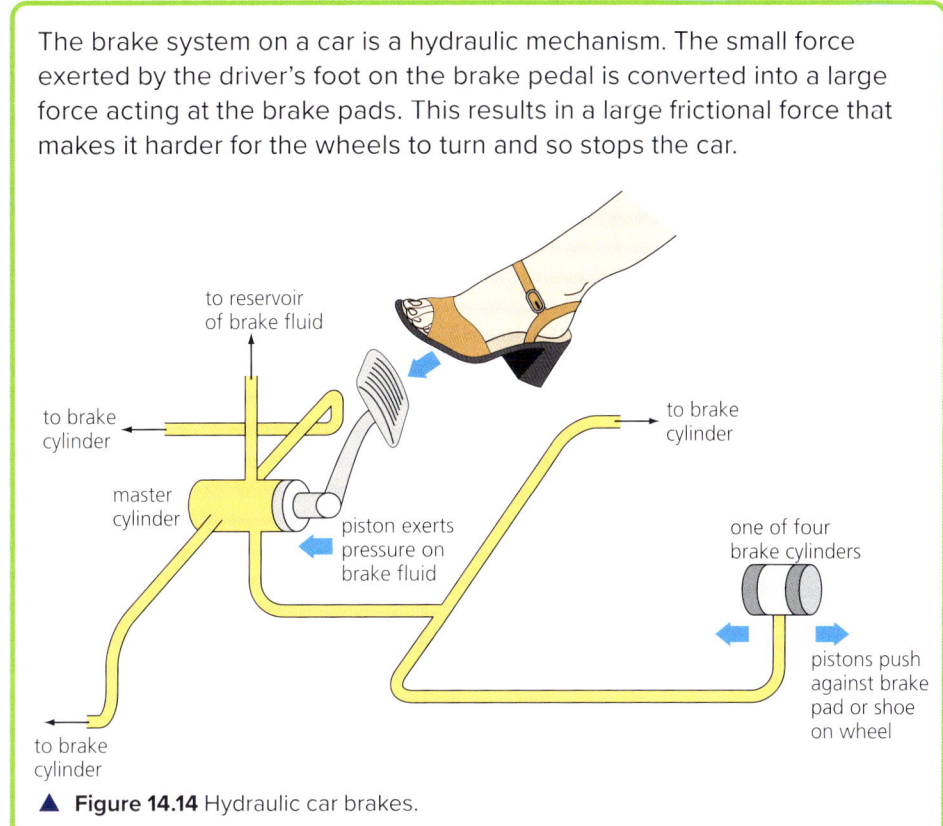

▲ **Figure 14.14** Hydraulic car brakes.

9 Why are hydraulic systems known as 'force multipliers'?

10 How important are hydraulic systems to the cars and trucks on the world's roads? Explain your answer.

Pressure in gases

Pressure of the atmosphere

The atmosphere is a mixture of gases. The molecules from which the gases are made move around, but are pulled down by the force of gravity exerted on them by the Earth. The atmosphere forms a layer of gases over the surface of the Earth that is about 1 000 km high. This creates a pressure of about 100 000 N/m² – equivalent to a mass of 10 tonnes on 1 m² – although this pressure lessens as you go up through the atmosphere.

You do not feel the weight of this layer of air above you pushing down, because the pressure it exerts acts in all directions, as it does in a liquid. Thus, the air around you is pushing in all directions on all parts of your body. You are not squashed because the pressure of the blood flowing through your circulatory system is strong enough to balance atmospheric pressure.

The atmosphere does not crush ordinary objects around us. For example, the pressure of the air pushing down on a tabletop is balanced by the pressure of the air underneath the table pushing upwards on the tabletop.

How can air pressure be tested with a glass of water?

You will need:

a glass, a piece of card, a jug of water, a large bowl, a small table, video camera, school yard (optional). Take care with the glass.

Hypothesis

Air pushes on everything around it. This pressure is strong enough to stop a small weight of water falling to the ground.

Prediction

If a volume of water is contained in a glass, covered by a card and turned upside down, the air pressure will push on the card and stop the water falling.

Investigation and recording data

1 Ask a friend to film you as you complete these steps in the investigation.
2 Take a glass and fill it to the top with water from the jug.
3 Place a piece of card over the top of the glass so that it covers it all.
4 Hold the card in place with your fingers and hold the glass over the bowl.
5 Keep your fingers in place and quickly turn the glass and card upside down over the bowl.
6 Carefully and slowly remove your fingers away from the card and look for it staying in place.
7 If the card does not stay in place the water will fall out into the bowl.

Examining the results

Look at the evidence provided by the film.

Conclusion

Compare the film evidence with the hypothesis and prediction and draw a conclusion.

Is your conclusion limited in some way? Explain your answer.

What improvements could be made? Explain the changes that you suggest.

Science extra: Ear popping

The middle part of the ear (shown in Figure 14.15) is normally filled with air at the same pressure as the air outside the body. The air pressure can adjust, because when you swallow, the Eustachian tubes in your throat open and air freely enters or leaves the middle ear.

For example, if the air pressure is greater outside the body and in the mouth, when you swallow, more air will enter the middle ear to raise the air pressure there.

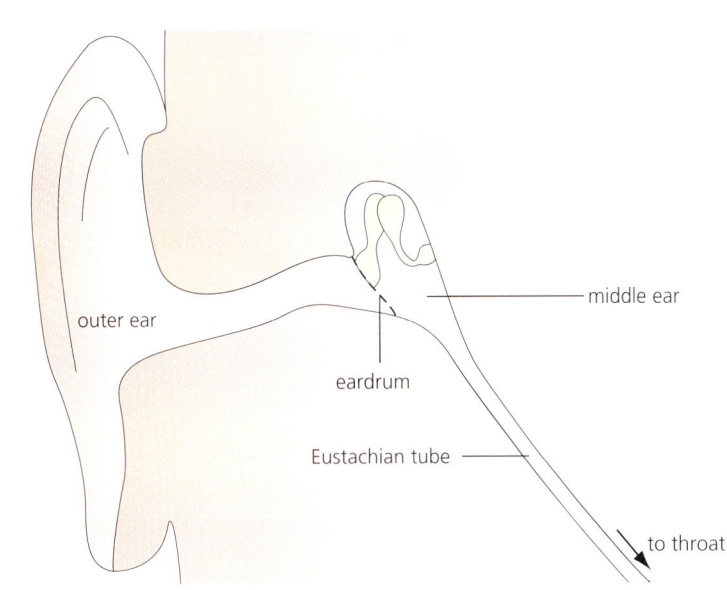

▲ **Figure 14.15** The ear and throat.

If you travel in a car that quickly climbs a steep hill, your ears sometimes 'pop'. This is because you are rising rapidly into the atmosphere where the air pressure is lower. The popping sensation is caused by the air pressure being lower in the throat and outside the body, and higher in the middle ear. The difference in pressure causes the eardrum to push outwards. When you swallow, the air pressure in your middle ear reaches the same pressure as the air in your throat and outside, and the eardrum moves quickly back – or 'pops' – into place.

Science extra: How a sucker sticks

When an arrow with a sucker on the end hits a target, the arrow stays in place due to air pressure. This can be explained by the particle theory. Before the sucker reaches the surface, it is surrounded by the same number of air particles pushing on all its surfaces. When the sucker presses against the target surface, some of the air particles are pushed out. There are fewer air particles pushing outwards on the sucker than there are air particles pushing inwards. This difference in pressure holds the sucker to the surface.

11 What happens if the air pressure in the throat and outside the body is lower than the air pressure in your middle ear when you swallow?

12 If you ride quickly down a hill on a bicycle, your eardrums are pushed in before they 'pop' back. Why is this?

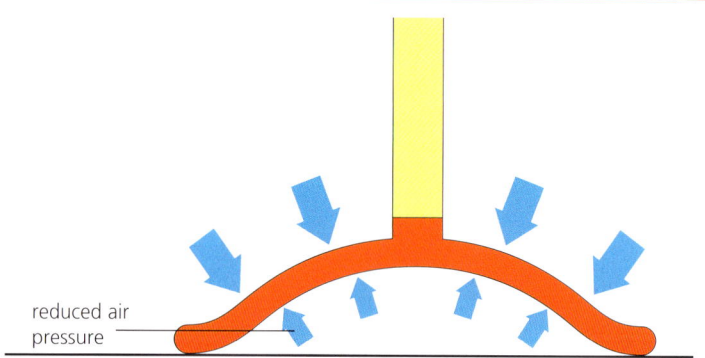

reduced air pressure

▲ **Figure 14.16** Side view through a sucker.

Science extra: Crushing a can

The strength of the air pressure in the atmosphere can be demonstrated by taking the air out of a can. This can be done in two ways:

Using steam

The can has a small quantity of water poured into it and is heated from below. As the water turns to steam it rises and pushes the air out of the top of the can. If the heat source is removed and the top of the can immediately closed, the remaining steam and water vapour in the can will condense, leaving only a small quantity of air in the can. This air has a much lower pressure than the air pressure outside the can, and the higher pressure crushes the can.

Using a vacuum pump

A **vacuum** pump can reduce the air pressure in containers. If one is used to remove air from a can, the can collapses due to the greater pressure of the air on the outside (Figure 14.17).

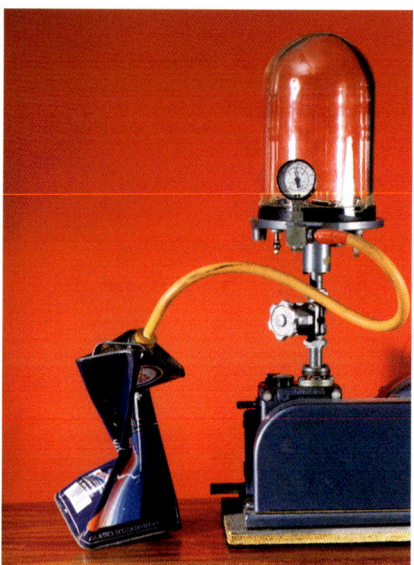

▲ **Figure 14.17** Removing air from this empty oil can has made it collapse.

Science in context

Hovercraft

Hovercraft are used around the world for transporting people and goods, and for rescue services around coasts. They can travel over shallow waters, where it is difficult for boats to go, and over mud flats into which people and vehicles would sink.

◄ **Figure 14.18**
A hovercraft.

A hovercraft uses the pressure of air to raise it from the ground. It does this by drawing air from above with powerful fans (see Figure 13.19). There is a skirt around the edge of the hovercraft, which prevents the air from escaping quickly, and the air pressure beneath the hovercraft increases. The upward pressure of the air trapped beneath the hovercraft lifts the hovercraft off the ground. The fans continue to spin to replace air that is lost from the edges of the skirt.

The cushion of air beneath the hovercraft reduces friction between it and the ground, and this cushion is also maintained when the hovercraft moves over water. The forward or backward thrust on the hovercraft is provided by propellers in the air above the hovercraft.

13 What are the advantages of using a hovercraft as a means of transport?

CHALLENGE YOURSELF

Can you make a hovercraft from an old CD, a plastic bottle top, some glue and a balloon?

Work out how you will use these items. Check your plan with your teacher, making sure your choice of glue is safe to use. If your plan is approved, make your model. Will it travel over the ground and over water?

propeller

air

fan

flexible skirt

▲ **Figure 14.19** Hovercrafts work by riding on a cushion of air above the ground or water's surface.

Diffusion

Diffusion is a process in which one substance spreads out through another. It occurs in gases and liquids.

▲ **Figure 14.20** The smell of street food cooking diffuses in the air around it.

Diffusion of gases

How can you tell if someone is cooking? It is by the smell, such as that of onions or spices, that reaches your nose. As someone cooks food, particles escape from the food's surfaces and enter the air. The food particles are knocked about by the particles in the air and move in a random way, moving away from each other and mixing up with the particles in the air. This mixing up is called **intermixing**. The particles from the food intermingle with the particles in the air. Over time, the particles from the food reach your nose by diffusion and you smell the food that is being cooked.

a
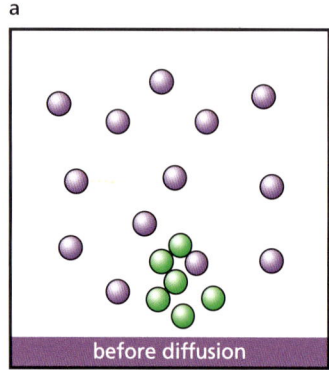
before diffusion

b
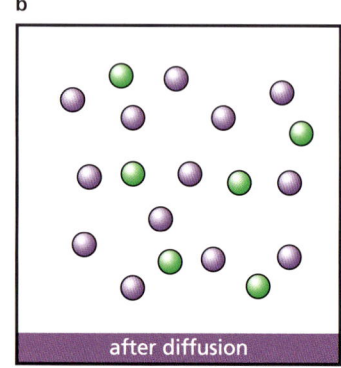
after diffusion

▲ **Figure 14.21a** Particles from food moving into the air; **b** particles of food spreading through the air by diffusion.

LET'S TALK

How can a stop-motion animation video be used to communicate the process of diffusion? If anyone in your group has used a stop-motion animation app, ask them to demonstrate it or, if someone is interested to find out, let them report back and assess the prospect of taking the project further. If the project moves to the next stage, decide on a storyboard, create it together, and then make the video.

Modelling diffusion

You will need:

a tray with a rim, and two sets of marbles, or clay or plastic balls in two different colours (such as red and blue).

Plan and investigation

1 Tip the tray slightly and set up a layer of one colour of marbles or balls. This can represent particles in the surface of the food being cooked.
2 Put five layers of the marbles or balls above them. This can represent the particles in the air above the food.
3 Photograph the scene.
4 Move one or two food particles into the air particles and photograph them again.
5 Repeat step 4 another four to six times.
6 Make a PowerPoint presentation using your photographs.

How fast does diffusion take place in gases?

You will need:

a perfume with a strong scent or a sample of hot food in closed containers, a dish, sticks with modelling clay on one end (for sticking to a bench, desk or table), a piece of brightly coloured card attached to the other end (to act as a flag) and a stop-clock.

Hypothesis

If a scent is released into the air and diffuses through it, it should be sensed by people at different distances over time.

Prediction

People nearer the place where the scent is released will smell it before people further away.

Be careful

Make sure that no one is allergic to the scent you have chosen and that the food does not contain nuts.

Investigation and recording data

1. Everyone in the class should have a stick (with a piece of card and clay on its ends) resting flat on their bench, desk or table.
2. One member of the class has a stop-clock and acts as an observer.
3. A member of the class takes the dish to one end of the room and places it on a table.
4. A small amount of scent or food is placed on the dish.
5. Everyone gets on with some written or reading work quietly so as not to disturb any air in the room.
6. As a person detects the smell of the scent or food, they should raise their stick and hold it in position with the clay. The observer then notes the time at which the scent arrived at that place; for example, 'desk 1, 10 seconds'.
7. The experiment continues until the people furthest from the dish detect (or appear unable to detect) the scent and the overall time of the diffusion process is observed and noted.

Examining the results

Look at the time the scent arrived at each desk and the distance of the desk from the scent source and look for a trend, pattern or anomalous result.

Conclusion

Compare your examination of the results with the hypothesis and prediction and draw a conclusion.

Is your conclusion limited in some way? Explain your answer.

What improvements could be made? Explain the changes that you suggest.

LET'S TALK
How can the data about the desk and the time of arrival of the scent be used with a tape measure to record the speed of diffusion across the room? In groups, work out a plan to find out, and, if approved by your teacher, try it.

Diffusion in liquids

Ink contains coloured particles. If you put a drop of ink in a beaker of water, it settles on the bottom, then spreads out by diffusion through the water and colours it, as shown in Figure 14.22.

14 Can the model you made in the activity on page 164 be used to show how ink diffuses through water? Explain your answer.

▲ **Figure 14.22** Black ink diffusing through a beaker of water.

How fast does diffusion take place in a liquid?

You will need:

a clear glass or plastic petri dish, a jug of water, a piece of graph paper, a coloured sweet (candy) and a stop-clock (or timer).

Hypothesis

If a coloured sweet is placed in water and the dye dissolves in the liquid, it should diffuse through the water.

Prediction

Make a prediction based on your knowledge and understanding of diffusion.

Investigation and recording data

1 Place the graph paper on the surface of a bench or table.
2 Place the petri dish over the graph paper. Make sure that you can see the squares through the bottom of the dish.
3 Pour in the water carefully until the dish is about two-thirds full.
4 Carefully lower the sweet into the centre of the dish and start the stop-clock or timer.
5 Look for the colour spreading out from the sweet and note the time as it passes over a square as it moves outwards. For example, 'square 1, 15 seconds', 'square 2, 25 seconds' and so on.
6 Note the size of the squares (such as 1 or 2 mm) and work out the speed of diffusion close to the sweet and, after a short time, further away from the sweet.

LET'S TALK

Heat is a form of energy which, if passed to particles, makes them move faster. How could you adapt the experiment to find out whether diffusion is affected by the heat of the water around the sweet? In groups, work out a plan to find out and, if approved by your teacher, try it.

Examining the results

Look at the time the coloured dye passed into the squares around the sweet. Look for a pattern, a trend and any anomalous results.

Conclusion

Compare your examination of the results with the hypothesis and your prediction and draw a conclusion.

Is your conclusion limited in some way? Explain your answer.

What improvements could be made? Explain the changes that you suggest.

Summary

✔ Pressure = force ÷ area
✔ Pressure is caused by the action of a force from a substance on an area.
✔ Pressures in gases and liquids are caused by the arrangement of their particles.
✔ Diffusion of gases and liquids can be described as the intermixing of substances by the movement of particles.

End of chapter questions

1 What is the equation that scientists use when they are studying pressure?
2 What is the pressure exerted on the ground by a cube which has a weight of 400 N and a side of
   a 2 m² b 4 m²?
3 When do you exert most pressure on the ground – when you are standing up or when you are lying down? Explain your answer.
4 Why do studs on your boots help you to play a game of hockey, rugby or soccer?
5 The pressure of the air at the Earth's surface is about 100 000 N/m². Why does it not squash you flat?
6 Imagine you were an engineer building a dam across a river. Should the base be thicker than the top? Explain your answer.
7 A bar of scented soap is placed in a room. At first no one can smell it, then later everyone can smell it. What has happened to make this change?
8 A shark can detect prey that has a wound which bleeds even if they are a large distance away. Explain how this can happen.

 Now you have completed Chapter 14, you may like to try the Chapter 14 online knowledge test if you are using the Boost eBook.

In this chapter you will learn:
- about reflection at a plane surface
- to use the law of reflection
- about objects with smooth surfaces (Science extra)
- about the speed of light (Science in context)
- about refraction of light at the boundary between air and glass or air and water
- about tricks of light (Science extra)
- that white light is made of many colours and that this can be shown through the dispersion of white light by using a prism
- about Newton's investigations with prisms (Science in context)
- about the rainbow (Science extra)
- how colours of light can be added, subtracted, absorbed and reflected.

Do you remember?

- Does light travel in straight or wavy lines?
- What is a ray diagram?
- A luminous object is something that gives out light. What is the luminous object which is providing light for you to read this book?
- Light can be reflected. What does this mean?
- How can you see an object that is not a light source?
- What happens to a ray of light when it is reflected from a mirror?
- Does a ray of light keep travelling in the same direction when it moves from air to water? Explain your answer.
- What do you understand by the term 'refraction'?

Light is a form of energy. In this chapter, we will look at how it is reflected and refracted, and investigate the colours of light.

Light-rays

Light leaves the surface of a **luminous object** in all directions but, if some of the light is made to pass through a hole, it can be seen to travel in straight lines.

For example, when sunlight shines through a small gap in the clouds, it forms broad sunbeams with straight edges (see Figure 15.1). The path of the light can be seen because some of it is reflected off dust in the atmosphere. Similarly, sunlight shining through a gap in the curtains of a dark room produces a beam of light which can be seen when the light reflects off the dust in the air of the room.

▲ **Figure 15.1** Although the Sun radiates light in all directions, the sides of sunbeams seem almost parallel because the Sun is a very distant luminous object.

Smaller lines of light, called **rays**, can be made by shining a lamp through slits in a piece of card.

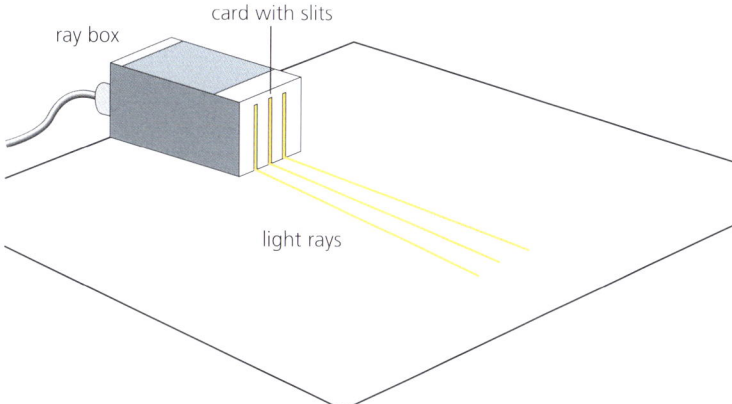

ray box

card with slits

light rays

▲ **Figure 15.2** Making rays of light.

Light-rays can be further investigated by letting them pass though a circular hole and strike a screen behind the hole. The device used to investigate light-rays in this way is called the **pinhole camera**. The ray diagram in Figure 15.3 on the next page shows light-rays passing from a light source to the screen in the pinhole camera.

1 How is the image of the light source on the screen different from the real light source?

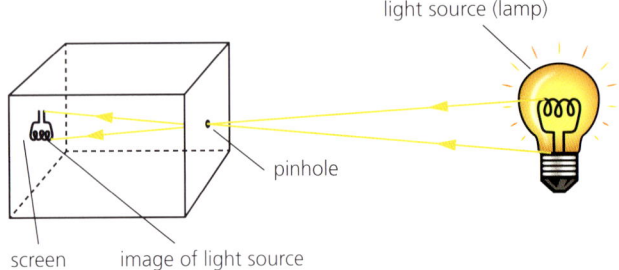

light source (lamp)

pinhole

screen   image of light source

▲ **Figure 15.3**

Light-rays that are reflected from a **non-luminous object**, such as a tree, can also pass through the hole in a pinhole camera and form an image on the screen. The ray diagram in Figure 15.4 shows light-rays passing from a tree to the screen in the pinhole camera.

2 Where do you think the light-rays reflected from the tree originally came from?

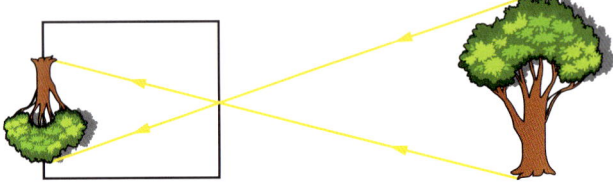

▲ **Figure 15.4**

If the pinhole camera is brought nearer to the tree, the light-rays from the tree travel as shown in Figure 15.5.

3 How does bringing the camera closer to the tree affect the image of the tree seen on the screen?

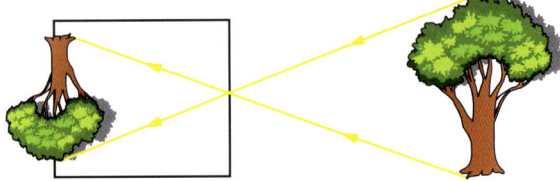

▲ **Figure 15.5**

4 Draw a ray diagram of the pinhole camera further away from the tree than in Figure 15.5.

Scientists sometimes make equipment to test the information they have read. Test the information you have just read by making a pinhole camera like the one shown in Figure 15.6, and then undertaking the suggested investigations.

pinhole

top

▲ **Figure 15.6** The parts of a pinhole camera.

Can you make and test a pinhole camera?

Work safely

You must never point the camera towards the Sun as it can permanently damage your eyes.

Take great care while using the needle.

You will need:

a cardboard box, aluminium foil, greaseproof or tracing paper, glue or sticky tape, a desk lamp or carbon filament lamp, a needle and a towel (optional).

Making a pinhole camera

1 Cut a circular hole in one end of the cardboard box, 15 mm in diameter, and cover this with aluminium foil (glued or taped to the box).
2 Cut a rectangular hole in the opposite side of the box, 50 mm × 80 mm. Cover this hole with greaseproof (or tracing) paper and glue it to the box.
3 Carefully use the needle to make a small hole in the centre of the foil.

Checking the performance of the pinhole camera

1 Set up a brightly lit object, such as a lamp.
2 Point the pinhole camera at the bright object, with the hole and foil pointing towards the object.
3 Look at the greaseproof paper screen for an image of the object.
4 If you cannot see an image, put the towel over the end of the box and hold it up with your hand to make it as dark as possible around the screen. The image should be visible now.

CHALLENGE YOURSELF

Create a PowerPoint presentation about making and using your pinhole camera. Compare it with other presentations and decide which provides the information most clearly.

Testing the information

1 Point your camera at the lamp and observe the image.
2 Point your camera at the tree and observe the image.
3 Move nearer to the tree and observe again.
4 Move further away from the tree and observe again, then move closer and further away and observe again each time.
5 Your observations should confirm the information you have read. If they do not, suggest improvements to your camera and technique that you could make.

CHALLENGE YOURSELF

When scientists make a piece of equipment, they may think of other investigations they could make with it. Here are two you might like to try.
1 How does the size of the hole affect the image on the screen of the pinhole camera? Note that you may have to replace the foil at the front of your camera before you try the next investigation.
2 Do many small holes make many images on the screen?

Reflection of light

A few terms are used in the study of light which make it easier for scientists to describe their investigations and ideas. In the study of **reflections**, the following terms are used:

- **Incident ray** – a light-ray that strikes a surface.
- **Reflected ray** – a light-ray that is reflected from a surface.
- **Normal** – a line that is at right angles (that is at 90°) to the surface that the incident ray strikes.
- **Angle of incidence** – the angle between the incident ray and the normal.
- **Angle of reflection** – the angle between the reflected ray and the normal.
- **Plane mirror** – a mirror with a flat surface.
- **Image** – the appearance of an object in a smooth, shiny surface. It is produced by light from the object being reflected by the surface.

The ways in which the incident ray, normal and reflected ray are represented diagrammatically are shown in Figure 15.7. The back surface of a mirror is usually shown as it is here, as a line with short lines at an angle to it. The way light-rays are reflected from a plane mirror can be investigated using the equipment shown in Figure 15.8.

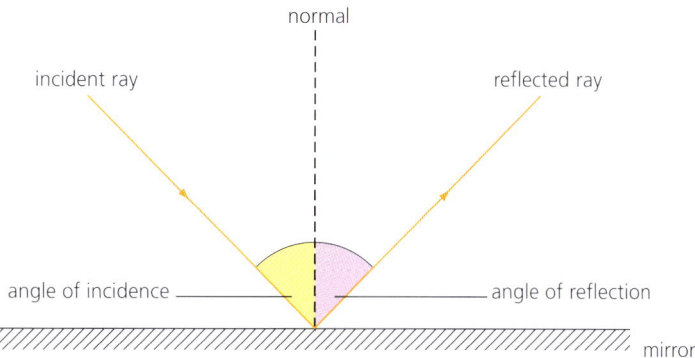

▲ **Figure 15.7** The reflection of light from a plane mirror.

The law of reflection

When scientists make a large number of experiments and observations on a particular topic, they may find that there is a trend or a pattern. If they do find a trend, they make a statement about the relationship, which is called a **law**. The angle between the incident ray and the normal is the same as the angle between the reflected ray and the normal. See if you can find a trend or pattern that leads to a law of reflection, by trying the following enquiry, which uses the equipment shown in Figure 15.8.

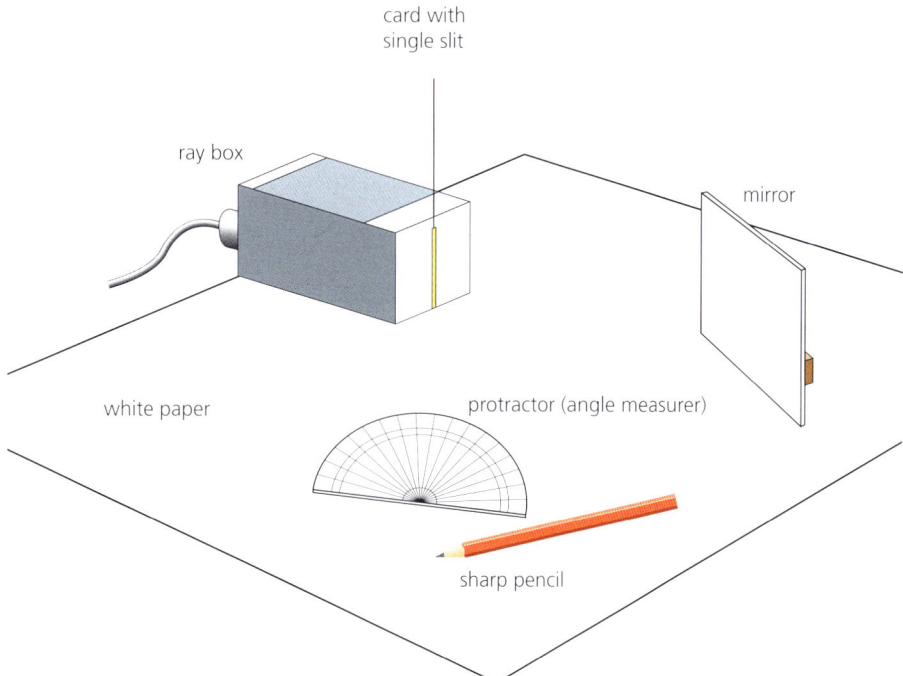

▲ **Figure 15.8** Equipment to investigate reflection from a plane mirror.

Scientists use patterns to guide them in order to create laws of science. In order to establish a law, they have to collect many examples.

5 Figure 15.9 shows three drawings made of the paths of incident and reflected rays in an experiment using the equipment shown in Figure 15.8. Use a protractor (angle measurer) to measure the angles of incidence and angles of reflection. What do these drawings tell you about the process of reflection?

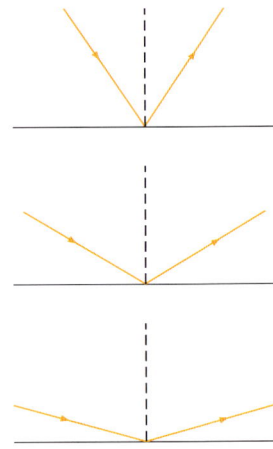

▲ Figure 15.9

Can you confirm the law of reflection?

You will need:

a ray box, a mirror, a protractor (angle measurer) and a piece of white paper.

Hypothesis
The law of reflection states that the angle between the incident ray and the normal is the same as the angle between the reflected ray and the normal.

Prediction
If the law is correct then when any ray of light is shone onto a mirror its angle of incidence and angle of reflection will be the same.

Plan, investigation and recording data
If a law of reflection can be established, the data from many experiments must be collected.

Plan an investigation to collect data that might support a law of reflection. In your plan, set out what you will do with the equipment you have been given and the table you will set up in which to record your results. If your teacher approves your plan, try it.

Examining the results
Compare the angles of reflection at different angles of incidence and look for a pattern in the data.

How can you tell if there are one or more anomalous results? If you find any in your data, identify them by marking them with an A.

Compare your data with the information you gathered in your answer to question 5.

Conclusion

From your evaluation, state the law of reflection you think you have discovered.

Is your conclusion limited in some way? Explain your answer.

What improvements could be made? Explain the changes that you suggest.

6 From the answers to Question 5, what do you think the law of reflection could say or state?

LET'S TALK

If you moved the mirror to a different angle on the paper and investigated the relationship between the angle of incidence and the angle of reflection, what do you think you would find? Would it support a law of reflection? Explain your answer. If you as a group think another investigation is required to answer the question, plan it and, if your teacher approves, try it. What do you find?

▲ **Figure 15.10** Light reflected from the smooth surface of a lake can produce an image in the water.

Science extra: Objects with smooth surfaces

Glass, still water and polished metal have very smooth surfaces. Light-rays striking their flat surfaces are reflected, as shown in Figure 15.11. Their angle of reflection is equal to their angle of incidence. When the reflected light reaches your eyes, you see an image (see Figure 15.10).

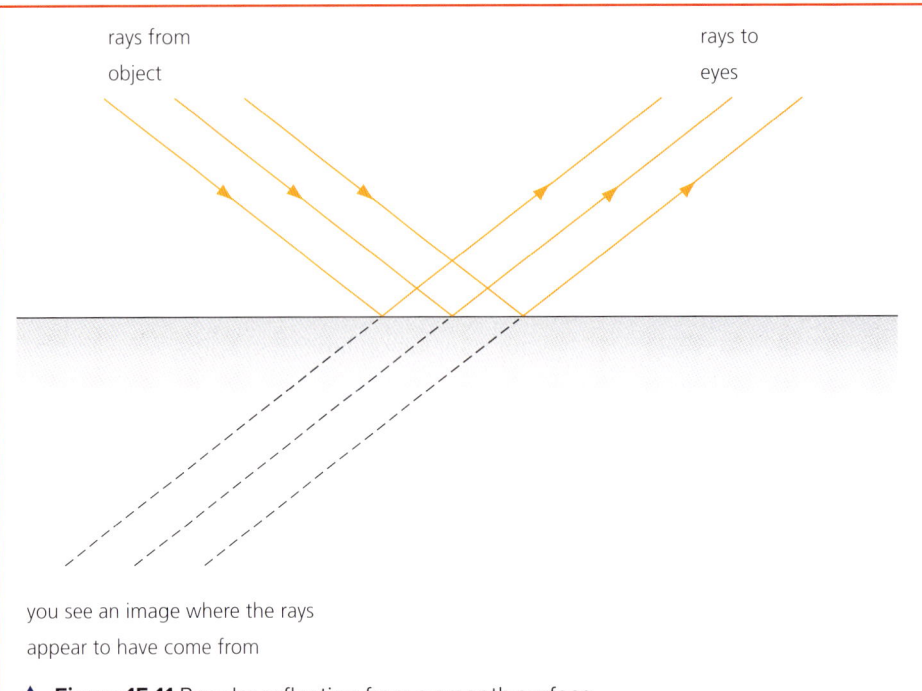

rays from object

rays to eyes

you see an image where the rays appear to have come from

▲ **Figure 15.11** Regular reflection from a smooth surface.

Science extra: Objects with rough surfaces

Most objects have rough surfaces. They may be very rough, like the surface of a woollen jumper, or they may be only slightly rough, like the surface of paper. When light-rays strike any of these surfaces, the rays are scattered in different directions (see Figure 15.12).

You see a jumper, or this page of your book, by the light scattered from its surface. You do not see your face in a piece of paper, because the reflection of light is irregular, so it does not form an image.

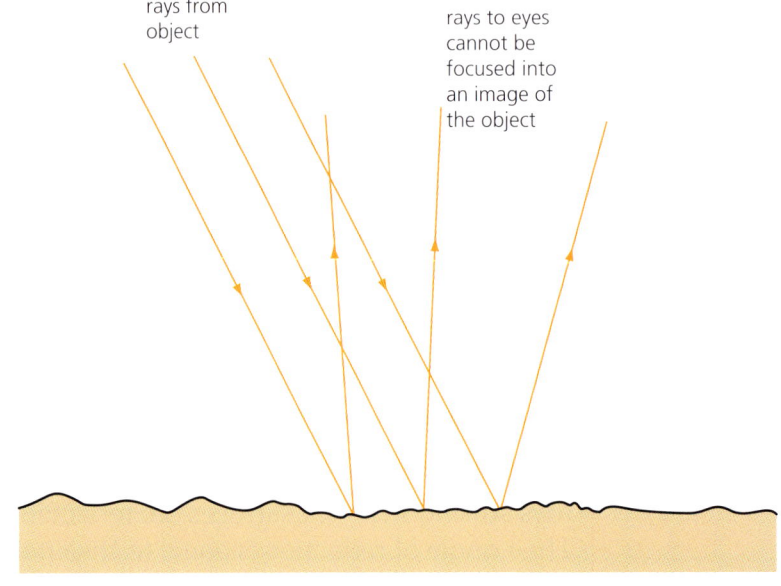

rays from object

rays to eyes cannot be focused into an image of the object

▲ **Figure 15.12** Light-rays are scattered by a rough surface.

The speed of light

The Ancient Greeks believed that light travelled at infinite speed, and this remained unchallenged until Ole Rømer (1644–1710), a Danish astronomer, observed the moons of Jupiter and studied how they travelled around the planet. When Jupiter was between the Earth and one of its moons, the moon could not be seen from the Earth and was said to be eclipsed by Jupiter. The four large moons move around Jupiter quite quickly and other scientists had found it possible to time them. When Rømer studied the eclipses more thoroughly, he discovered that they appeared to occur earlier when the Earth was nearer Jupiter in its orbit than when it was further away (see Figure 15.13).

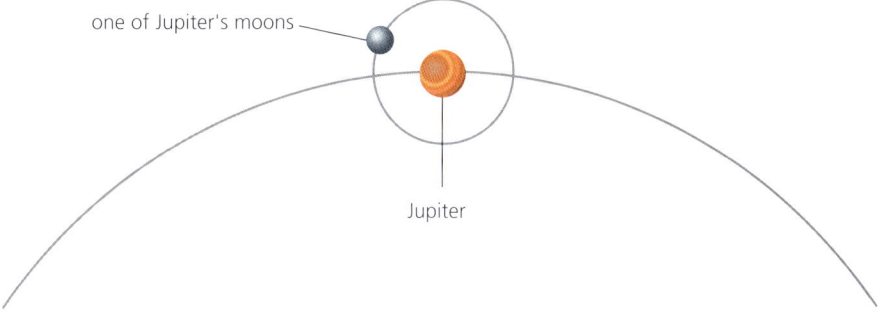

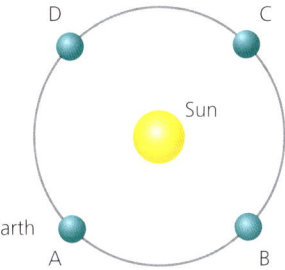

▲ **Figure 15.13** The positions of the Earth in its orbit when Rømer made his observations.

Rømer did not believe that the moons speeded up at different times of the year. He believed the difference was due to light having a finite speed, and that it took longer to reach the Earth when the Earth was at points A and B than when it was at points C and D. By taking measurements and making calculations, Rømer deduced a **speed of light** which showed that light took 11 minutes to get from the Sun to the Earth.

James Bradley (1693–1762), an English astronomer, studied the position of the stars at different times of year as the Earth moved in its orbit. From his studies he calculated the speed of light. His results showed that light took 8 minutes and 11 seconds to travel from the Sun to the Earth.

7 What evidence about the speed of light had Rømer to work with when making his studies?

8 What two pieces of evidence about Jupiter's moons did Rømer use to plan his investigation?

9 What did Rømer's measurements show?

10 What creative thought did Rømer have to explain his measurements?

11 How accurate was Bradley's calculation of the time it takes light to reach the Earth from the Sun? Explain your answer.

12 How accurate was Fizeau's value for the speed of light compared to the current-day value? Explain your answer.

In 1849, Armand Fizeau (1819–1896), a French physicist, made an instrument which measured the speed of light from a candle placed 9 kilometres away. He made many measurements and calculated that light travels at a speed of 314 262 944 meters per second (m/s).

Many other scientists refined Fizeau's work by making more complicated pieces of equipment, and today the speed of light has been measured as 299 992 460 m/s in a vacuum, slightly slower in air and even slower in water and glass. The speed of light in air is often rounded to 300 000 000 m/s and the average (mean) time taken for light to travel from the Sun to the Earth has been measured as 8 minutes and 17 seconds.

▲ **Figure 15.14** Armand Fizeau.

Refraction of light

If a ray of light is shone on the side of a glass block, as shown in Figure 15.15a, the ray passes straight through but, if the block is tilted, the ray of light follows the path shown in Figure 15.15b. This 'bending' of the light-ray is called **refraction**. The angle that the refracted ray makes with the normal is called the angle of refraction (see Figure 15.16).

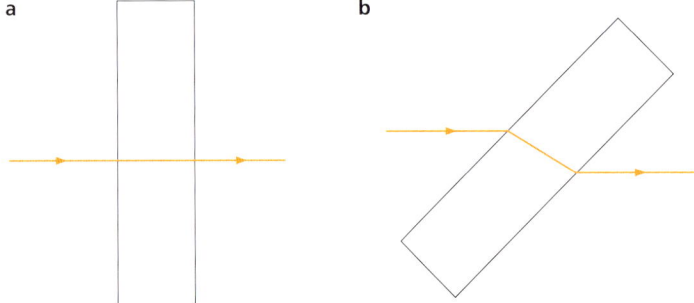

▲ **Figure 15.15** Light is refracted if the incident ray is not at 90° to the surface of the transparent material.

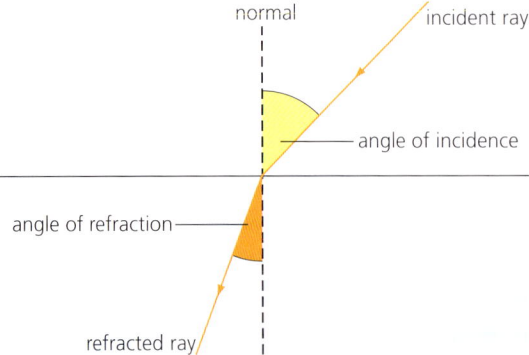

▲ **Figure 15.16** The angle of incidence and the angle of refraction.

The refraction of light as it passes from one transparent substance or 'medium' to another is due to the change in the speed of the light. Light travels at different speeds in different media. For example, it travels at almost 300 million m/s in air but only 200 million m/s in glass. If the light slows down when it moves from one medium to the other, the ray bends towards the normal. If the light speeds up as it passes from one medium to the next, the ray bends away from the normal.

Science extra: Tricks of light

Light speeds up as it leaves the water's surface and enters the air. A light-ray appears to have come from a different direction than that of the path it actually travelled (see Figure 15.17). The refraction of the light-rays makes the bottom of a swimming pool seem closer to the water's surface than it really is. It also makes streams and rivers seem shallower than they really are, and this fact must be considered by anyone thinking of wading across a seemingly shallow stretch of water. The refracted light from a straw in a glass of water makes the straw appear to be bent.

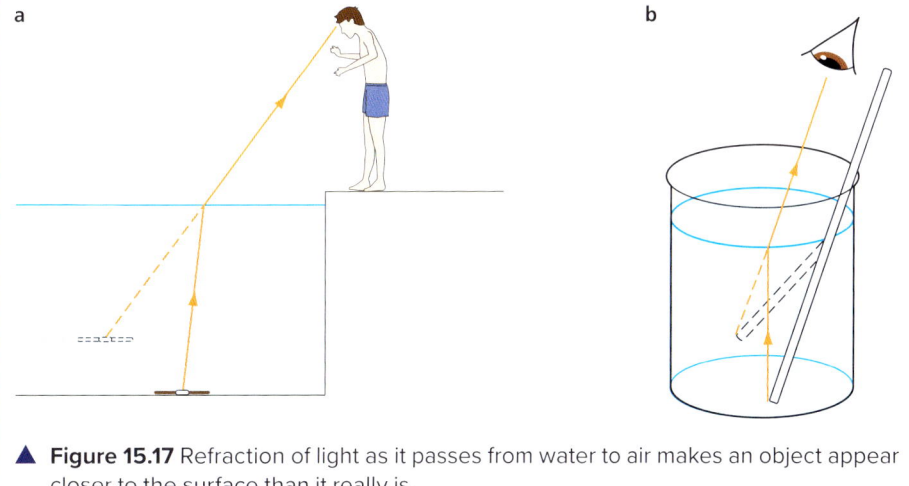

13 How is the reflection of a light-ray from a plane mirror (see pages 172–3) different from the refraction of a light-ray as it enters a piece of glass?

▲ **Figure 15.17** Refraction of light as it passes from water to air makes an object appear closer to the surface than it really is.

Science extra: Can you make a coin appear?

You will need:

an opaque bowl, a coin, a jug of water, a helper and a camera.

Hypothesis

If a coin is placed in water, the light coming from it will be refracted when it enters the air and will make the coin appear to be in a different place.

Prediction

Use your knowledge and understanding of refraction to predict how your view of the coin may change.

Investigation and recording data

1 Put the bowl on a table and set up a coin in its centre.
2 Step back from the bowl until the coin disappears under the rim of the bowl.
3 Start filming and focus on the rim of the bowl and the inside of the bowl where the water will be placed.
4 Ask a helper to slowly and carefully fill up the bowl with water.
5 Keep filming and look for the coin to appear. If and when it does, stop filming.

Examining the results

Play back the film and let others see it.

Conclusion

Use your knowledge of refraction to explain your observations and your film.

White light and the prism

A triangular **prism** is a glass or plastic block with a triangular cross-section. When a ray of sunlight is shone through a prism at certain angles of incidence, and its path is stopped by a white screen, a range of colours, called a **spectrum**, can be seen on the screen.

This splitting up of white light into different colours is called **dispersion**. It occurs because the different colours of light travel at very slightly different speeds and are refracted at slightly different angles by the prism. This makes them spread out, or disperse, as Figure 15.18 shows.

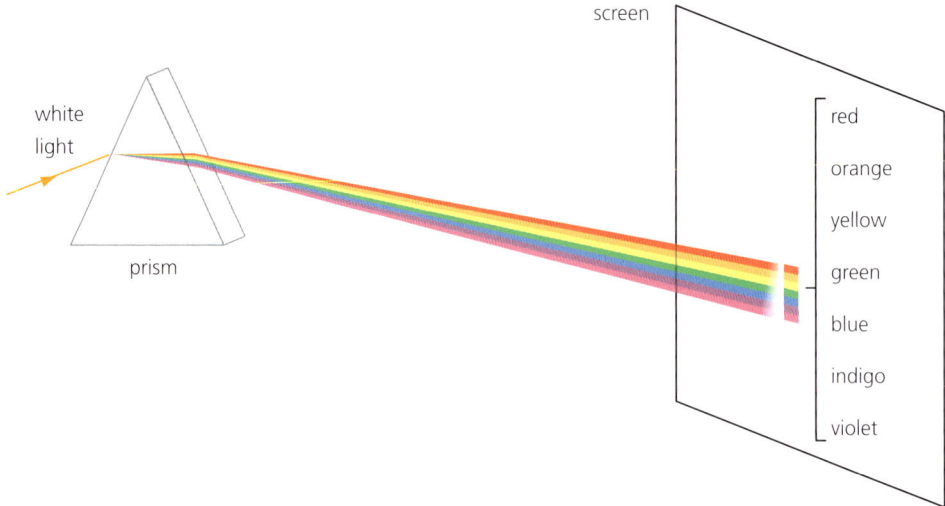

14 What happens to white light when it passes through a prism?

▲ **Figure 15.18** White light passing through a prism is split up into its constituent colours, forming a spectrum.

Newton's investigations with prisms

Isaac Newton (1642–1727) was an English scientist who made many scientific investigations. His investigation of light began when he bought a prism at a fair. He put it in a dark room and shone sunlight onto it from a hole in a window shutter. The light entered the prism and dispersed to form a spectrum of colours on a screen, as Figure 15.19 shows.

▲ **Figure 15.19** Newton dispersing light with a glass prism.

Newton thought that it was possible that the prism could have produced the colours just by the light shining through it. To check this idea, he got a second prism and set it up behind the first, but the opposite way round. When the rays of the separate colours of light passed through the second prism, they formed a beam of white light, as Figure 15.20 shows.

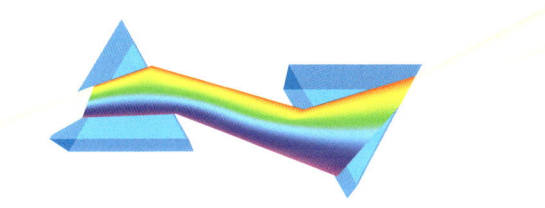

▲ **Figure 15.20** Newton's second experiment with prisms.

This proved that white light is made up of light of different colours that were dispersed by the first prism.

CHALLENGE YOURSELF

Explain how the light passed through the two prisms by using the words 'refracted' and 'dispersed'. You may use the word 'refracted' a few times.

Science extra: The rainbow

If you stand with your back to the Sun when it is raining, or if you look into a spray of water from a fountain or a hose, you may see a rainbow. It is produced by the refraction, dispersion and reflection of light-rays in a raindrop. Figure 15.21 shows the path of a light-ray and how the colours in it spread out to form the order of colours – the spectrum – seen in a rainbow.

Sometimes a second, weaker rainbow is seen above the first because two reflections occur in each droplet. In the second rainbow, the order of colours is reversed.

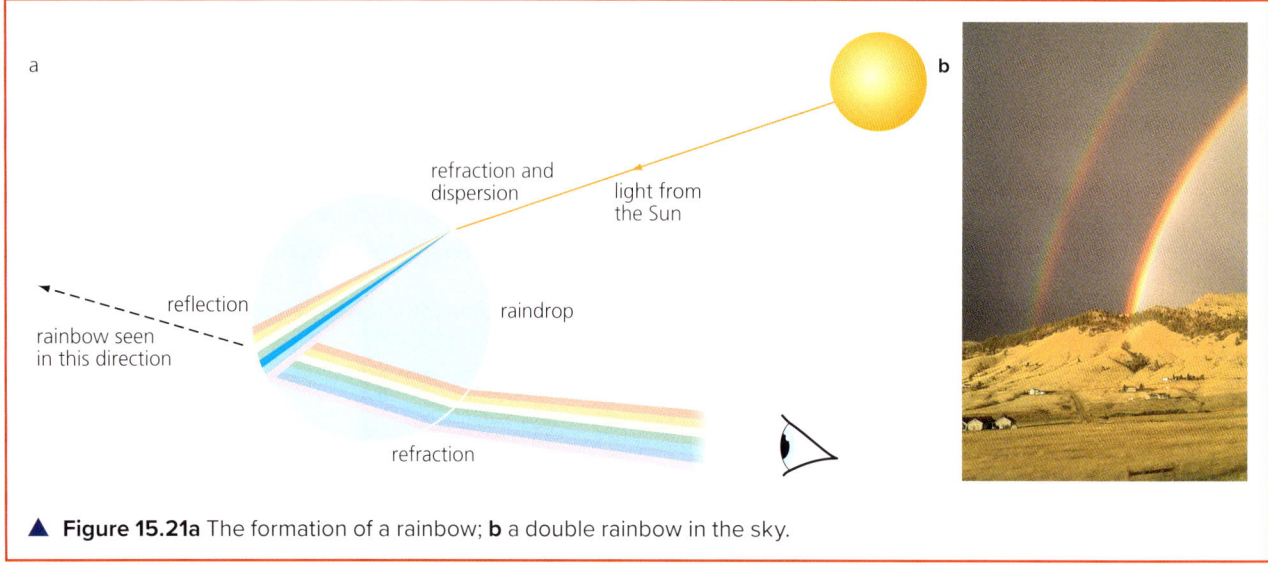

▲ **Figure 15.21a** The formation of a rainbow; **b** a double rainbow in the sky.

Science extra: Modelling a raindrop in a rainbow

You will need:

a transparent glass full of water, a source of white light (for example, a flashlight), a piece of white card and a dark room.

Process

1 Set up the glass of water on the card in a darkened room and shine the flashlight on it.
2 Move the flashlight around until you see a spectrum form on the card. You may have to look hard to see it.
3 Comment on your model by identifying the analogy, and assess how good you were at modelling a raindrop.

Colour

Absorbing and reflecting colours

When a ray of sunlight strikes the surface of an object, all the different colours in it may be reflected, or they may all be absorbed. If all the colours are reflected, the object appears white; if all the colours are absorbed, the object appears black.

Most objects, however, absorb some colours and reflect others. For example, healthy grass reflects mainly green and absorbs other colours.

15 Name some everyday objects which:
  a reflect all the colours in sunlight
  b absorb all the colours in sunlight.

Filtering colours

Some sweets are wrapped in transparent, coloured coverings and you may have looked thorough them to see the world as green or red. The coverings were acting as light filters. In science experiments, sheets of coloured glass or plastic are used to filter the colours in light. The colour is produced by dyes which are put into the filter when it is being made.

16 What happens to white light when it is shone on a
a blue filter
b green filter?

Each filter absorbs all the colours in white light except one, and this single colour passes through. This means that when white light shines on a red filter, all the colours in the light are absorbed except red, and this passes out of the filter as red light. All the colours in white light have been subtracted from it by the filter, except red, which passes through.

Something that you almost certainly did not do with the coloured sweet wrappers was to shine a light through them. The following investigation allows you to shine lights through coloured filters and see what happens when the different colours meet and mix.

What happens when red, blue and green light mix?

You will need:

three colour filters (red, blue and green), three flashlights and a piece of white card.

Plan, investigation and recording data
Construct a plan by answering these questions.
1 How will you make beams of coloured light using the filters and flashlights?
2 When you are shining the light beams on the white card, how will you make your test fair?
3 You should begin by shining two beams onto the card at once.
4 If you find that a new colour has formed, describe it. The following words may be helpful: pink, light blue, yellow.
5 What happens when you shine all three beams onto the white card? You may need someone to hold a flashlight for you when you do this.
6 How will you record your observations in step 5?

Examining the results
Look at the colours produced when the three colours of light are shone together onto the white card.

Conclusion
Draw conclusions from your observations.

Addition and subtraction of coloured light

Colour addition

When different coloured lights are combined, it is found that all the colours can be made from different combinations of just three colours. They are red, green and blue and they are called the primary colours of light.

When beams of the three primary colours are shone onto a white screen so that they overlap, they produce three **secondary colours** of light and white light (see Figure 15.22).

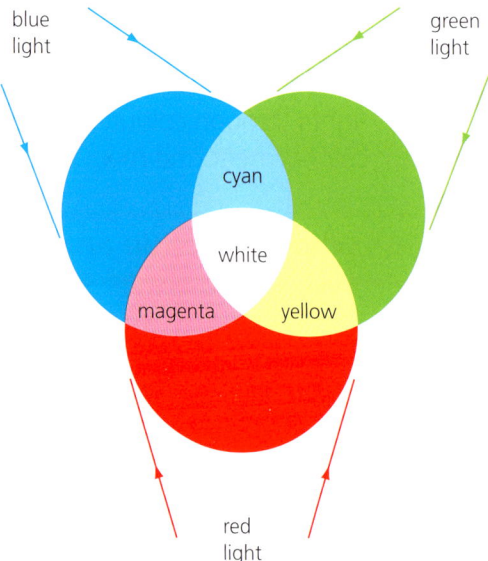

blue light

green light

cyan

white

magenta   yellow

red light

▲ **Figure 15.22** Overlapping beams of the primary colours form the secondary colours.

When light is brought together like this, the wavelengths of red, blue and green light come together to produce the colours you see in Figure 15.22, and when all three colours are added together, white light is produced. This combining of the three colours of light is called colour **addition** and the mixing of light is called **additive colour mixing**.

17 The colours on a television or computer screen are made by three different colours of substances called **phosphors**. They glow to release their colour of light.

What do you think the colours of the phosphors are? Explain your answer.

18 Look at Figure 15.22. Which primary colours overlap to produce:
   a yellow
   b magenta
   c cyan
   d white light?

Colour subtraction

When white light passes through a blue filter, all the colours in the white light are absorbed, except blue light, which passes through the filter. All the colours in white light have been **subtracted** from the light, leaving the filter. This results in blue light passing through the filter and making the filter appear blue as Figure 15.23 shows.

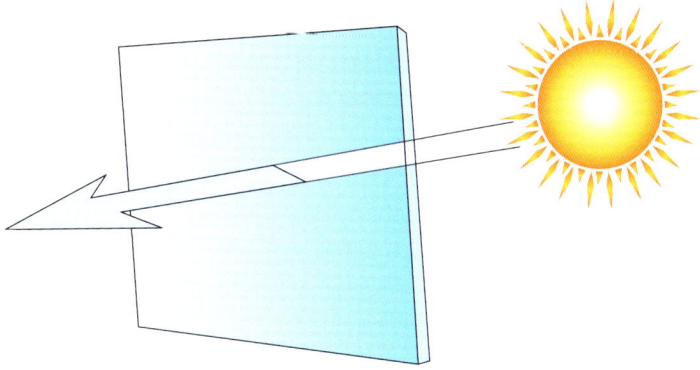

▲ **Figure 15.23** Colour subtraction.

Summary

✔ We can use the incident ray, normal and reflected ray to show the reflection of light on a plane surface.
✔ The law of reflection states that the angle of incidence and angle of reflection are the same.
✔ Science in context: The speed of light is rounded to 300 000 000 m/s.
✔ Light refracts at the boundary between air and glass or air and water.
✔ White light is made of many colours and this can be shown through the dispersion of white light by using a prism.
✔ Isaac Newton's investigations with prisms confirmed that white light was made up of light of different colours.
✔ Colours of light can be added, subtracted, absorbed and reflected.

End of chapter questions

1 What happens when a ray of light is reflected from a smooth flat surface?
2 What does the law of reflection say?
3 If a light-ray has an angle of incidence of 35°, what is its angle of reflection?
4 How do you measure an angle of incidence?
5 How does the path of a light-ray change when it moves from one transparent material to another?
6 What is the refraction of light and where does it occur?
7 Why is the speed of light important when light is refracted by a prism?
8 Explain what happens when white light is shone on a green filter.
9 What happens when a beam of red light is mixed with a beam of blue light?

 Now you have completed Chapter 15, you may like to try the Chapter 15 online knowledge test if you are using the Boost eBook.

In this chapter you will learn:

- about magnetic fields
- that a magnetic field surrounds a magnet and exerts a force on other magnetic fields
- that the Earth has a magnetic field because its core acts as a magnet
- about early discoveries about magnetism (Science in context)
- about factors that change the strength of an electromagnet
- about some of the ways that electromagnets can be used
- how to make an electromagnet.

Do you remember?

- A magnet has two poles, one at each end. What are they called?
- What happens to two magnets when they attract each other?
- What happens to two magnets when they repel each other?
- What is the difference between a magnet and magnetic material?
- Do the magnets have to be touching each other to attract or repel each other? Describe what happens.
- Does a magnet have to be touching a magnetic material to hold onto it? Describe what happens.
- What does the following statement mean? 'Different poles attract and similar poles repel.'

1 How are these magnets holding things onto the fridge door?

DID YOU KNOW?

There are three metallic elements which have strong magnetic properties. They are iron, nickel and cobalt. The alloy of iron, called steel, has magnetic properties too.

▲ **Figure 16.1** Magnets are not only used to hold messages on fridge doors, but a magnetic strip in the fridge door is also used to hold it closed.

▲ **Figure 16.2** Magnetite is a naturally occurring magnet.

It is thought that the word 'magnet' comes from the name of the ancient country of Magnesia, which is now part of Turkey. In this region, large numbers of black stones were found which had the power to draw pieces of iron to them. The black stone became known as lodestone or leading stone, because of the way it could be used to find directions, and it eventually came to be used in the compass. Today, it is known as the mineral magnetite, and it has been found in many countries.

CHALLENGE YOURSELF

The force of gravity pulls everything down to the centre of the Earth, but what happens if you have a magnetic force pull against it? Try to find the answer by performing the following experiment.

Can a magnetic force make something hover in the air?

You will need:

a camera, a box, a piece of thread, a ruler, some sticky tape, two pieces of modelling clay, a disc/fridge magnet and a steel paper clip.

Hypothesis

The force of gravity pulls things down, but a magnetic force acting on a small object might be able to pull it up against gravity.

Prediction

If the object is small and the magnet is near but not touching, it can pull up the object against gravity.

Plan, investigation and recording data

Use the items stated and Figure 16.3 to work out a plan to investigate the question then try it.

If you make the paper clip float, take a photograph of it or make a short film.

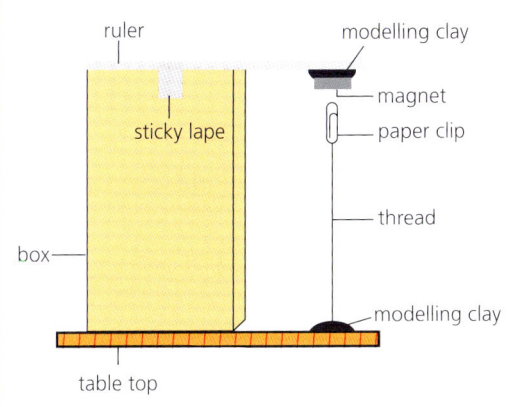

▲ **Figure 16.3** The equipment set up.

Examining the results

Did the paper clip fall or hover?

Conclusion

Did the evidence of the enquiry support or contradict the hypothesis?

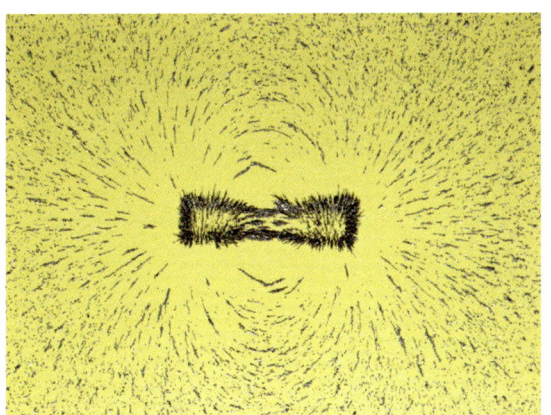

▲ **Figure 16.4** The magnetic field pattern of a bar magnet as shown by iron filings.

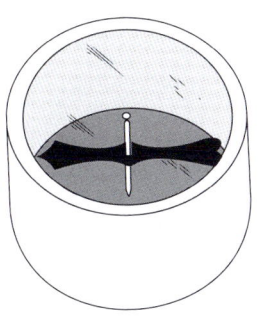

▲ **Figure 16.5** A plotting compass.

The magnetic field

There is a region all around the magnet in which the pull of the magnetic force from the magnet acts on magnetic materials. This region is called the **magnetic field**.

The field around a magnet can be shown by using a piece of card and iron filings. The card is laid over the magnet and the iron filings are sprinkled over the paper.

Each iron filing has such a small mass that it can be moved by the magnetic force of the magnet if the paper is gently tapped. The iron filings line up as shown in Figure 16.4. The pattern made by the iron filings is called the magnetic field pattern.

If you look closely at Figure 16.4, you can see that some of the iron filings close to the magnet form lines that arch over from one end of the magnet to the other. These lines are called **lines of force**.

You may already know about a compass and how to use one, and starting on page 195 there is more information about them. For now, you need to know that there is a small compass called a **plotting compass** which can be used to investigate magnetic fields.

Can a line of force be found with a plotting compass?

You will need:

a bar magnet, a plotting compass, a sheet of white paper or card and a pencil.

Hypothesis

A compass needle is a magnet, and will react in a magnetic field like the iron filings, which are made of magnetic materials.

Prediction

The changing position of the compass needle as it moves from one end of a magnet to another can be used to plot a line of force.

Investigation and recording data

1   Place the bar magnet in the centre of the card.
2   Place the plotting compass close to one pole of the magnet and note the direction in which the needle points.
3   Take the pencil and mark a point on the paper where the needle is pointing.
4   Move the compass just ahead of the point you have made and note the direction the needle is pointing now.
5   Take the pencil again and mark a point on the paper where the needle is pointing.
6   Repeat steps 4 and 5 as many times as you need to reach the other pole of the magnet.

Examining the results

Compare the line you have made on the card with Figure 16.4.

Conclusion

Draw a conclusion to answer the enquiry question.

How can a magnetic field be found with a plotting compass?

You will need:

a bar magnet, a plotting compass, a sheet of white paper or card and a pencil.

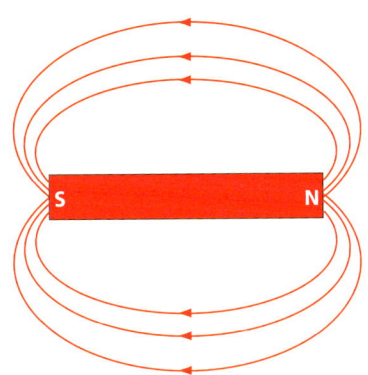

▲ **Figure 16.6** The magnetic field pattern around a bar magnet.

Hypothesis

A compass needle is a magnet, and will react in a magnetic field like the iron filings, which are made of magnetic materials. If it is placed at different distances from the magnet, it should allow you to plot a number of lines of force.

Prediction

The plotting compass may show different lines of force, as shown in Figure 16.6.

Investigation and recording data

1 Place the plotting compass at the corner of one end of the magnet and plot a line of force as in the previous enquiry.
2 Move the plotting compass a little more towards the centre of one end of the magnet and try to plot another line of force.
3 Move the plotting compass to the centre of one end of the magnet and try to plot another line of force.
4 Repeat steps 1–3, starting at the other corner of the same end of the magnet.

Examining the results

Compare the results of your plotting with Figure 16.6 and the magnetic field shown by the iron filings.

Conclusion

Draw a conclusion to answer the enquiry question.

Is your conclusion limited in some way? Explain your answer.

What improvements could be made? Explain the changes that you suggest.

2 Figure 16.7 shows how iron filings spread out when in contact with the end of a bar magnet. Make a drawing of how you think the field lines are arranged all around the magnet.

▲ **Figure 16.7**

Lines of force and magnetic strength

We have seen that the distance between the lines of force varies along the length of the magnet. This observation can be used to set up another enquiry to see if there is a link between the distance between the lines of force and the magnetic strength.

Is there a link between lines of force and magnetic strength?

You will need:

a dish of steel paper clips, a bar magnet and a camera.

Hypothesis

The lines of force are close together near the poles and further apart between the poles. The magnetic strength is greatest where the lines of force are close together.

Prediction

Magnetic objects will collect in greatest numbers at the poles.

Investigation and recording data

1 Hold the bar magnet horizontally.
2 Slowly lower the bar magnet towards the dish of paper clips.
3 Let the bar magnet touch the paper clips in the dish, then slowly raise the magnet.
4 Observe and record the result with a photograph.

Examining the results

Examine and note the positions of the paper clips on the bar magnet. Note the places where most paper clips are found. Compare the positions of the paper clips with the positions of the lines of force you found in the previous enquiry, and look for links between them, such as a trend.

Conclusion

Draw a conclusion by examining the results and comparing them with the prediction.

Is your conclusion limited in some way? Explain your answer.

What improvements could be made? Explain the changes that you suggest.

The magnetic fields when two magnets meet

We have seen how the magnetic force from a magnet forms a magnetic field which can be investigated to show a magnetic field pattern as seen in Figure 16.4. We can now take this knowledge forwards to see what happens to the magnetic fields and their field patterns when two magnets are brought together.

What are the magnetic fields like when two magnets meet?

You will need:

two bar magnets, a plotting compass, two sheets of white paper or card and a pencil.

Hypothesis

When the poles of two magnets are brought together, the field pattern they create can be plotted by plotting compasses.

Prediction

The field pattern between the poles will depend on whether the poles are the same or opposite.

Investigation and recording data

1   On one sheet of paper, make drawings of the three possible horizontal arrangements of the two bar magnets.
2   Draw lines of force to predict the field pattern between each pair of magnetic poles.
3   Set up the pair of magnets in the first arrangement on a sheet of card.
4   Use the plotting compass to mark out lines of force and make a magnetic field pattern.
5   Repeat steps 3 and 4 with the next two arrangements of the magnets.

Examining the results

Compare the field patterns you have constructed with those in the given prediction.

Conclusion

Draw a conclusion from your examination of the results.

Is your conclusion limited in some way? Explain your answer.

What improvements could be made? Explain the changes that you suggest.

Looking at how magnetic field patterns can change

Each magnet has its own magnetic field pattern, but when another magnet is brought close to it, the field patterns of both magnets change. This is due to the way the magnetic fields interact. If two similar poles are bought together, a certain pattern develops between the two magnets, but if two opposite poles are brought together, a different field pattern is produced.

3 Figure 16.8a and b show two magnetic field patterns when two magnets are brought together. From your experiment on the field patterns of two magnets, which one shows when the two poles are
   a similar
   b opposite?

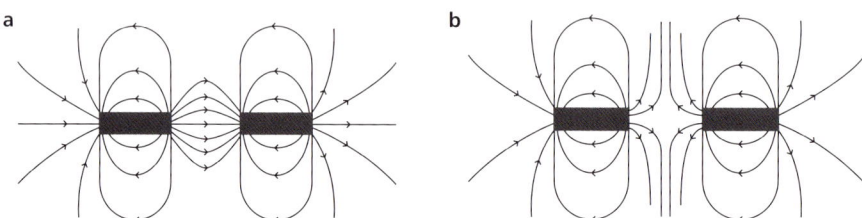

▲ **Figure 16.8a and b**

The Earth's magnetic field

At the centre of the Earth is the Earth's core. It is made from iron and nickel and is divided into two parts: the **inner core**, made of solid metal, and the **outer core**, made of liquid metal. As the Earth spins, the two parts of the core move at different speeds and this is thought to generate the magnetic field around the Earth, making it seem as though the Earth has a large bar magnet inside it.

The Earth spins on its **axis**, which is an imaginary line that runs through the centre of the planet. The ends of the line are called the geographic north and south poles. Their positions on the surface of the Earth are fixed.

Magnetic north – towards which the free north pole of a magnet points – is not at the same place as the geographic north pole (see Figure 16.9) and it changes position slightly every year.

4 a Look at the field pattern around the Earth in Figure 16.9. Which pole of the imaginary bar magnet inside the Earth coincides with magnetic north?

b Copy Figure 16.9. On your diagram, draw a bar magnet inside the Earth and label its poles. Also label the position of the magnetic south pole on the Earth's surface.

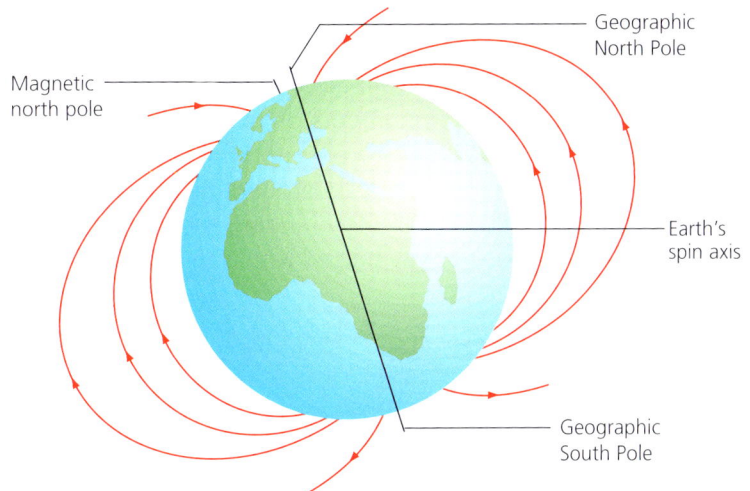

Geographic North Pole

Magnetic north pole

Earth's spin axis

Geographic South Pole

▲ **Figure 16.9** The Earth's geographic and magnetic poles do not coincide.

The magnetic north pole originally got its name because it is the place to which the north poles of bar magnets point. In reality it is the Earth's south magnetic pole because it attracts the north poles of magnets. Similarly, the magnetic south pole is really the Earth's north magnetic pole because it attracts the south poles of bar magnets. However, for most purposes the old, incorrect names for the magnetic poles are still used.

Science in context

Early discoveries about magnetism

Probably the first use of a magnet in direction-finding was in the practice of Feng Shui by the Ancient Chinese civilisation. They used a device called a luopan (see Figure 16.10), which contained a magnet to find a south-pointing direction. They then read off a scale around the magnet to decide on the final position of building foundations.

◀ **Figure 16.10** A luopan.

The first evidence of the magnet being studied scientifically for navigation is found in the writings of Chinese scientist Shen Kuo (1031–1095). He performed experiments on magnetic needles, described how magnets pointed north and south, and how other directions (east and west) could be found using a scale around the magnet. The knowledge of using magnetite for direction-finding is believed to have slowly passed to other countries as they traded with one another.

◀ **Figure 16.11** Chinese scientist Shen Kuo.

Petrus de Peregrinus (also known as Peter the Pilgrim) was a French engineer who lived in the thirteenth century. He experimented on the way magnets could attract and repel each other and how they could point north and south. He believed that the magnet pointed to the outer sphere of the heavens. Compasses at that time were made by floating a magnetic needle on water, but Peregrinus showed that attaching the needle to a pivot made the compass easier to use.

William Gilbert (1544–1603) was an English scientist and doctor to Queen Elizabeth I. He made many experiments on magnets and disproved beliefs such as 'garlic destroys magnetism' and 'rubbing a diamond on a piece of iron makes the iron into a magnet'.

Gilbert suspended a magnetic needle so that it could move both horizontally and vertically, and discovered that the needle also dipped as it pointed north–south. He extended his investigation by using a model of the Earth made out of a sphere of lodestone (magnetite). He put a compass with a pivot at different places on the surface of his model Earth, and showed that the dip varied with the position of the compass on the sphere, just as it did with compasses at different places on the surface of the Earth.

D.ʳ WILL.ᴹ GILBERT,
Phyſician to Q.ⁿ Elizabeth.
From an Original Picture in the Bodleian Library Oxford.

◀ **Figure 16.12** English scientist William Gilbert.

From this investigation, Gilbert described the Earth as behaving as if it contained a huge magnet. In time, from this early work on magnetism, a compass was developed, like the one shown in Figure 16.13, to help people in every country find their way around the world.

◀ **Figure 16.13** A nineteenth-century mariner's compass.

5 How do we know about the work of Shen Kuo?

6 How was knowledge of magnetism and direction-finding spread to other parts of the world?

7 What evidence from the work of Peregrinus did Gilbert use when he did his scientific modelling?

8 Gilbert could have used two methods to show that rubbing iron with a diamond did not make a magnet. What do you think these were?

9 How do you think Gilbert tested whether garlic destroys magnetism?

10 When Gilbert saw that the magnet not only pointed but also dipped north–south, what creative thought do you think entered his mind?

11 What further creative thought do you think Gilbert had to test his idea?

12 How did Gilbert's explanation of the reason for magnets pointing north–south differ from the explanation given by Peregrinus?

The link between magnetism and electricity

Hans Christian Ørsted (1777–1851) was a Danish physicist who studied electricity. In one of his experiments, he was passing an electric current along a wire from a battery when he noticed the movement of a compass needle which had been left near the wire. This chance observation led to many discoveries about how magnetism and electricity are linked together and it has many modern applications.

When an electric current passes through a wire, it generates a magnetic field around the wire. A compass can be placed at different positions on a card around the wire (see Figure 16.14) and the lines of force can be plotted.

When the current flows up through the card, the field shown in Figure 16.15a is produced. When the current flows down through the card, the field shown in Figure 16.15b is produced.

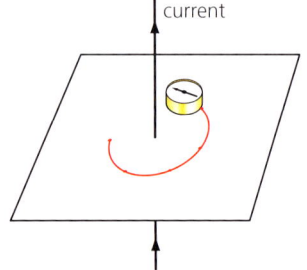

▲ **Figure 16.14** Plotting magnetic field lines around a current-carrying wire.

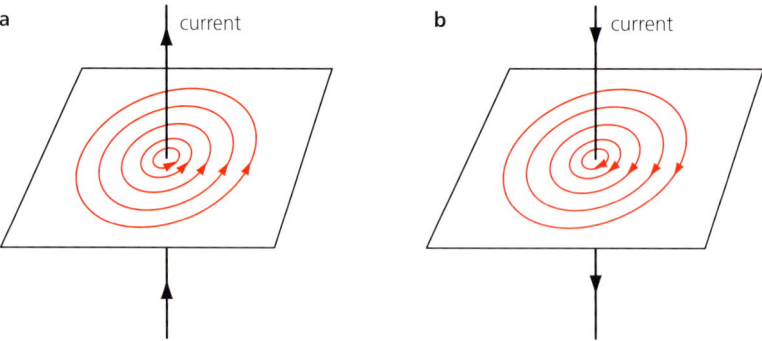

▲ **Figure 16.15** The magnetic field around a current-carrying wire.

Lines of force on diagrams of magnetic fields show not only the direction of the field as given by a plotting compass, but also the strength of the field in different places. The lines of force are close together where the field is strong and further apart where the field is weaker.

If the wire is made into a coil and connected into a circuit, a magnetic field is produced around the coil as shown in Figure 16.16.

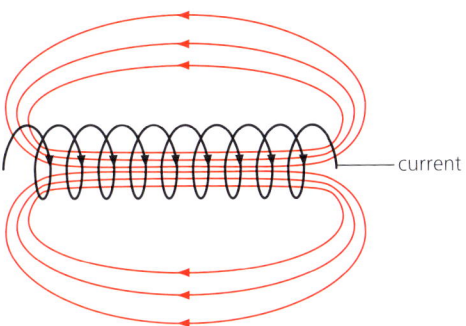

current

▲ **Figure 16.16** The magnetic field around a current-carrying coil.

13 How are the magnetic fields in Figures 16.15a and b different?

14 How does the strength of a magnetic field around a wire vary?

15 Compare the magnetic field around a bar magnet (Figure 16.6) with that produced by a current in a wire coil (see Figure 16.16).

The electromagnet

An **electromagnet** is made from a coil of wire surrounding a piece of iron. When a current flows through the coil, magnetism is induced in the iron and the coil and iron form a strong electromagnet. When the current is switched off, the electromagnet completely loses its magnetism straightaway. This device, which can instantly become a magnet and then instantly lose its magnetism, has many uses. For example, a large electromagnet is used in a scrapyard to move the steel bodies of cars (see Figure 16.17).

16 Describe how you think an electromagnet can be used to make a stack of scrapped cars, three cars high.

▲ **Figure 16.17** An electromagnet in use in a scrapyard.

How can you make and test an electromagnet?

Making an electromagnet

You will need:

a 5 cm iron nail, 3 m of thin, plastic covered copper wire and a ruler.

1   Measure 25 cm from one end of the copper wire, then wind the next 50 cm of wire around the nail.

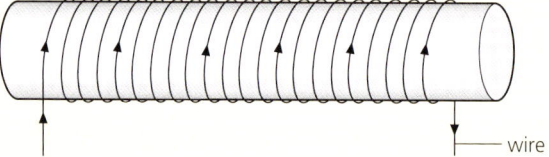

wire

▲ **Figure 16.18**

2   The turns of the wire should form a single layer over the nail – they should be close together and all wound in the same direction.

How to test an electromagnet

You will need:

your electromagnet, a switch, three cells, three wires, a clamp and stand.

1   Set up your electromagnet on the clamp and stand so that it is a few centimetres above the bench or table top.
2   Attach one end of a wire to one end of the wire of your electromagnet.
3   Attach the other end of the wire to a cell.
4   Attach the end of another wire to the other end of the wire of your electromagnet.
5   Attach the other end of this wire to one terminal of a switch.
6   Connect the cell and switch with the third wire to complete the circuit.
7   Hold some paper clips under one end of your electromagnet, switch on the circuit and record how many paper clips are attracted to your electromagnet.

Does the number of cells in the circuit affect the power of your electomagnet?

1   Make a plan to answer the question, and if your teacher approves, try it.
2   What does your investigation show?

Does the number of coils in an electromagnet affect its magnetic power?

Plan, investigation and recording data

1 Make a plan to investigate the enquiry question.
2 State the equipment you will need and how you will investigate and record data.
3 If your teacher approves, try it.

Examining the results

Examine your data and make comparisons.

Conclusion

Draw a conclusion to answer the question.

Does changing the size of the current affect the strength of an electromagnet?

You will need:

all the equipment set up as in the previous enquiry.

Hypothesis

The size of the current of electricity affects the strength of the electromagnet.

Prediction

Make your own prediction from the hypothesis.

Plan and investigation

Using the information from your test, work out a plan for how to investigate the electromagnet's strength between 0.2 and 2 amps of current. If your teacher approves, try it.

Record

Note the number of paper clips held at each value of current, and present the data in a table.

Analysis and evaluation

Examine your data and, if you detect a pattern, state it.

Conclusion

Compare your evaluation with your prediction and make a conclusion.

CHALLENGE YOURSELF

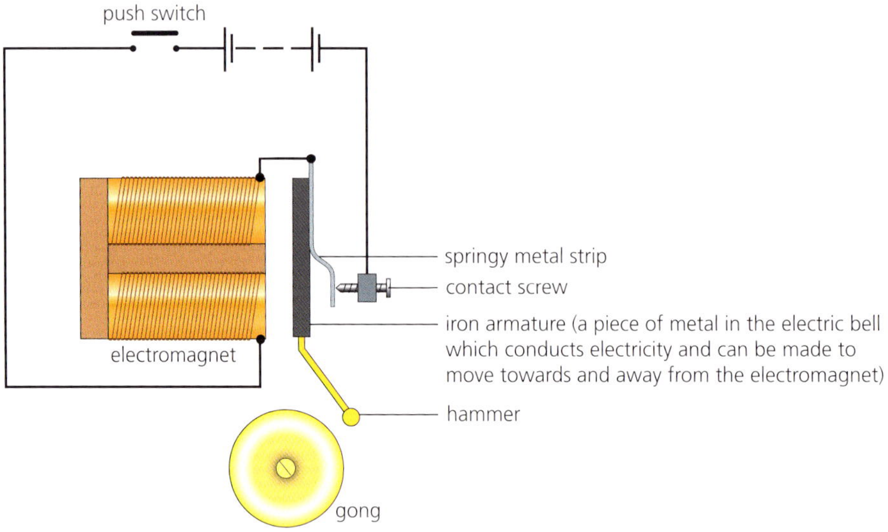

▲ **Figure 16.19** The circuit of an electric bell.

1   When the switch is pushed, the current passes though the electromagnet. The iron armature is on a spring-metal strip and can move.
    a   What happens to it now?
    b   What happens to the hammer?
2   When the armature moves, a gap forms between it and the contact screw. What happens next
    a   to the electromagnet
    b   to the hammer?
3   When you push the switch, the bell goes 'ding, ding, ding'. Why does it do this and not make just one 'ding'?

Summary

✔ The region around a magnet in which the pull of the magnetic force from the magnet acts on magnetic materials is called the magnetic field.
✔ A magnetic field surrounds a magnet and exerts a force on other magnetic fields.
✔ The Earth has a magnetic field because its core acts as a magnet.
✔ An electromagnet is made from a coil of wire surrounding a piece of iron and can be used in many ways, including as a scrapyard magnet.

End of chapter questions

1 Where would you find a magnetic field?

2 If you put iron filings round a metal bar that you thought was a magnet, how could you tell if it was magnetic?

3 What is a line of force and where would you find it?

4 If you bring the south pole of one magnet and the north pole of another magnet close together, how do the magnetic fields change?

5 What happens to an electromagnet when you switch it

    a on   **b** off?

6 State two things you could change about an electromagnet and its circuit to make it stronger.

7 What happens inside Earth to make a magnetic field around it?

8 Why can you use a compass to find your direction on the surface of the Earth?

Now you have completed Chapter 16, you may like to try the Chapter 16 online knowledge test if you are using the Boost eBook.

17 Using the Earth's resources

In this chapter you will learn:
- how humans use resources
- about renewable energy resources, such as wind, tidal and solar power
- about the environmental problems of setting up a wind farm (Science in context)
- about non-renewable energy resources, such as fossil fuels
- about our impact on ecosystems by using resources (Science in context).

Do you remember?

- Name six materials that we use in our daily lives.
- Name three useful materials that are found in rocks.
- Name five forms of energy.
- Which of the following statements about energy are true and which are false?
  - Energy can be made.
  - Energy can be destroyed.
  - Energy can be transferred.

We cannot survive without using the Earth's **resources**. Two resources we have used since our earliest times are resources that provide us with energy and resources that provide us with materials. Each of these resources can be divided into two – renewable sources and non-renewable resources. Renewable resources can be replaced in a short time and so are not used up. Non-renewable resources cannot be replaced and will eventually be used up.

Renewable material resources

A renewable material is a material that naturally replaces itself. Wood is a very good example of a **renewable material resource**. You chop down a tree, plant some seeds which grow into new trees and in time you have more trees to provide you with wood for your needs.

◀ **Figure 17.1** Forestries organise the harvesting of trees and plant new ones to replace them.

Other common renewable materials are cotton, wool and bamboo.

DID YOU KNOW?

Bamboo is the fastest-growing renewable material and has many uses, including being used for making roads, bridges, homes, clothes and even schools.

▶ **Figure 17.2** Bamboo is used in construction around the world.

Bioplastics

Other renewable materials include straw, sawdust, cornstarch, and fats and oils produced by plants. These renewable materials are used to make **bioplastics**, which have the advantage over plastics in that they break down more quickly in the environment and cause much less pollution. An example of a bioplastic that you can make easily in the laboratory is casein, which is made from milk.

1 Is your food a renewable material? Explain your answer.

How can you make plastic from milk?

Work safely

Make sure that the equipment has cooled down before clearing it away.

If carrying a burning splint from a Bunsen burner or spirit burner that is already lit, take great care not to bump into anyone.

Hands should be washed thoroughly after this experiment.

You will need:

a beaker, a Bunsen burner or spirit burner, a tripod, gauze, a measuring cylinder, some whole full-fat milk (not semi-skimmed), some vinegar, a stirring rod, a clock or timer, a kitchen sieve, a spatula or spoon and some paper towels.

Investigation and recording data

1 Measure out $240\,cm^3$ of milk and pour it into the beaker.
2 Assemble the tripod and gauze and put the Bunsen or spirit burner underneath them.
3 Place the beaker of milk on the gauze, light the Bunsen or spirit burner and heat the milk to just below its boiling point.
4 Measure out $59\,cm^3$ of vinegar and pour it into the hot milk.

5 Stir the milk and vinegar for 1 minute and look for lumps developing in the mixture.
6 Filter the mixture through a kitchen sieve and use a spatula to lift out the lumps and place them on a paper towel.
7 Gather the lumps together on the paper towel and squash them to remove as much water as possible.
8 Describe the properties of your plastic.

Conclusion

Does your investigation answer the enquiry question?

Keep your bioplastic for the decay test in the Challenge yourself after this enquiry.

It is claimed that bioplastics have the properties of many plastics and can be used in place of them. It is also claimed that when the bioplastics are thrown away, they break down into simple materials and do not remain and pollute the environment.

CHALLENGE YOURSELF

The decay test

1 Do bioplastics decay faster than oil-based plastics?
2 Do bioplastics all decay at the same rate?

Plan an investigation to answer these two questions and, if your teacher approves, try it. Keep a record of your investigation over a period of weeks, then analyse and evaluate your data before drawing conclusions. How limited is your conclusion? Explain your answer.

Non-renewable material resources

During the formation of the rocky crust of the planet, metal ores and minerals formed. Many of these materials have been found to be useful to us and are extracted from the rock, purified and set to a wide variety of functions, from the electronic circuits in cell phones to the bodies of cars and aircraft. All of these are **non-renewable material resources**. They cannot be regenerated quite quickly like renewable materials. This means that when they have been removed from the crust through mining in one place, other places must be found for mining to continue. Most of the plastic that we use today has come from chemical reactions on fossil fuels. Fossil fuels are also a **non-renewable energy resource** (see pages 212–3).

2 One way to recycle some materials would be to use them as packaging and wrapping for presents. Suggest some materials that you could use (other than old wrapping paper) and think about how you could make them into an unusual wrapping for the next presents you give.

LET'S TALK

How long do you keep a metal drinks can, a plastic drinks bottle and a bag for clothes? What could you do to increase their use to you? Could they be used for something else, like the tyres being made into shoes, as shown in Figure 17.3? How would using the items more than once help to conserve materials? How would this help the environment?

◀ **Figure 17.3** This man is making shoes made from old car tyres.

When you have finished with these materials, they can all be recycled along with other items, such as batteries and cartridges for computer printer inks. How often do people think about recycling, and how often do they do it when they have finished with a material? What kind of poster could be made to promote recycling in your school and neighbourhood?

Renewable energy resources

Wind as a source of energy

The kinetic energy of the moving air particles in the wind is transferred to the **turbine** blades and makes them turn. The kinetic energy of the turbine blade is transferred to a magnet surrounded by a coil of wire in the generator. As the magnet spins, its energy is transferred to the coil of wire in the form of electrical energy.

▲ **Figure 17.4** A wind turbine farm in India.

Making a turbine is an application of science in engineering, and you can use your engineering skills to make a turbine which will perform a simple energy transfer – the kinetic energy from the wind to a turning shaft – and use this energy to exert a pulling force against gravity.

Can you make a wind turbine that can perform some work?

You will need:

internet access, a stopwatch or timer, materials for making the turbine and assessing its performance.

Plan, investigation and recording data

1 Use the internet to look at wind turbines or windmills made by school children.
2 Select a design that you could develop. Your turbine will need a shaft at the back which is turned as the turbine turns.
3 Make a drawing of your turbine, showing how it is supported and the shaft at the back to which you will attach a small mass by a thread. When the turbine turns, this mass will be raised to show that the turbine can do some work.
4 Show your drawing to your teacher and, if approved, make your turbine and test it.
5 Record how fast the mass is raised.

Examining the results

Examine the turbine after it has raised the mass and look for signs of weakness that may have developed as it performed its task.

Conclusion

After examining the performance and condition of the turbine, state how well you have answered the enquiry question.

Is your conclusion limited in some way? Explain your answer.

What improvements could be made? Explain the changes that you suggest.

CHALLENGE YOURSELF
Does your turbine perform better than others made in your class? Organise a competition to find out who can make the fastest and most powerful turbine.

Science in context

The environmental problems of setting up a wind farm

Setting up a wind farm of wind turbines in a landscape is often controversial – that is, some people think it is good idea, and others think it harms the environment in some way.

Factors to consider
Here are some of the factors to consider in setting up a wind farm.

Obstacles such as buildings can cause turbulence in the air when the wind passes over them and, if the turbine is sited near a building, it may not get a steady flow of wind to keep it turning smoothly.

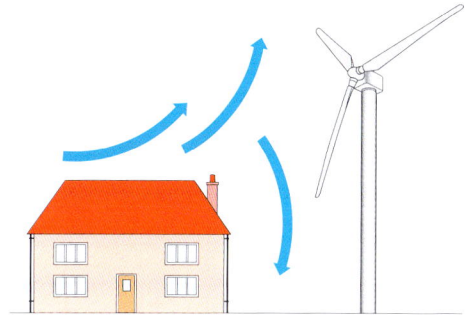

▲ **Figure 17.5** A house near a turbine.

When wind blows against a cliff face, turbulence develops at the top. If a turbine is sited at a clifftop, it will not turn smoothly.

A hill that rises steadily to a rounded hilltop, then sinks slowly on the other side, is a good site for a turbine. It should be set up at the top of the hill. The wind here will move smoothly over the hilltop, like the air on an aircraft wing, and keep the turbine turning steadily.

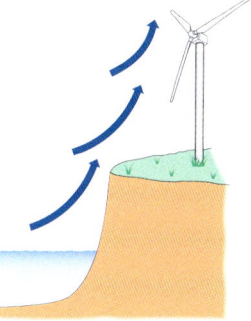

▲ **Figure 17.6** A turbine on a clifftop.

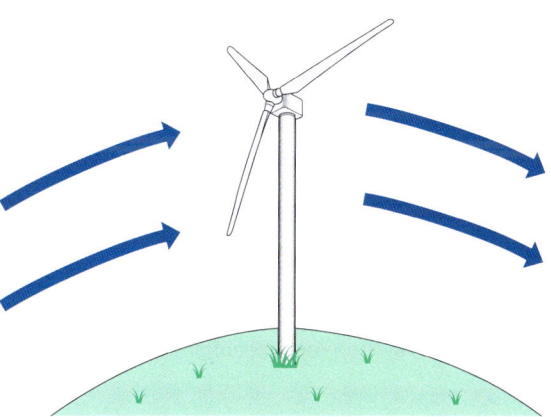

▲ **Figure 17.7** A turbine on a hilltop.

Trying to solve the environmental problem
Imagine you live in a town on the coast. Behind the town are two hills. One hill has no mountains behind it and no one lives there. It is covered in small plants and low bushes and is the home of many wild birds. There are a few paths across it for hikers to use. The second hill has mountains behind it and there is a town at its foot.

3 Why is it important for the turbine to turn smoothly and not keep stopping and starting?

CHALLENGE YOURSELF

Imagine that you are a science reporter who is going to write an article giving a balanced view, setting out the advantages and disadvantages of different sites for the wind farm. Research your article by reading more about renewable and non-renewable energy. Write your article and see if your friends think it is balanced or if you seem to favour one site more than another.

The council (the elected people who run the town) decide that they could cut down on the use of coal and oil to generate electricity by setting up a wind farm. The problem is where to build it. It could be on the hill without the mountains behind it, where it could be easily seen from the town. It could be on the hill with the mountains behind it, where the turbines would be less easily seen, but their noise might disturb the people in the town. Lastly, it could be built out at sea, where it would not be seen, but there would be more expense in bringing the electricity back to the town.

When issues such as this (selecting a site for a wind farm) arise, the media (newspapers, radio and television) run reports and articles about them. They often take one view in order to try and influence the decisions of the people affected by the issue.

Tides as a source of energy

Energy from tides

The energy stored in the water at high **tide** can be used to generate electricity. A dam is built across the **estuary** and the rising water of the tide passes through pipes in which turbines turn. This movement is passed to a generator and electricity is produced. When the tide falls, the water passes through the pipes in the opposite direction and turns the turbines again, so more electricity is generated.

▲ **Figure 17.8** A hydro-electric dam generates electricity from the tides.

4   The kinetic energy in the moving water in a river is a source of renewable energy. If you just placed turbines in the river to generate electricity for a power station, how reliable would the electricity supply be? Explain your answer.

5   How is building a dam for the power station a better choice than leaving the turbines in the river?

6   Where should you place the turbines in the dam to use the longest-lasting, fastest-moving jets of water passing through the dam wall?

LET'S TALK

When a dam is built for power stations using river water, a large reservoir is built behind it. This means that the land behind the dam must be flooded and the people living there must be moved. Figure 17.9 shows a village which could be flooded.

◄ **Figure 17.9** A village in a valley that could be flooded to make a reservoir.

Imagine you have to explain to the leader of a village that the people will have to move. You have to make the movement of villagers out of the area as peaceful as possible. You can offer them incentives to move, such as money and land, but you must try and move them as cheaply as possible. You may like to develop this activity into someone taking the role of the village leader and others taking the role of villagers, while you and a few others take the roles of people representing the company who are building the dam and power station.

CHALLENGE YOURSELF

Waves as a source of energy

Waves move up and down as they pass by. Machines have been developed to convert the up-and-down motion of the wave into a circular motion, which can turn the shaft of a turbine and generate electricity. The machines need to be moored in places where the waves are frequent, but not strong enough to damage or destroy them (see Figure 17.10).

◄ **Figure 17.10** This machine converts the energy in the up-and-down motion of passing waves into a turning motion required to make an electric generator work.

1 Draw an energy transfer diagram for a generator using the energy in a wave.
2 What expenses would have to be paid to set up machines to generate electricity from waves and transport the electricity to the shore?

Solar energy

We can use two forms of solar energy as sources of renewable energy. Heat energy can be trapped by solar panels and can then be used to heat buildings (Figure 17.11). Light energy can also be converted into electrical energy by solar cells (Figure 17.12).

◄ **Figure 17.11** These roofs in Sweden are fitted with solar panels.

◄ **Figure 17.12** Arrays of solar cells project from the International Space Station to collect some of the Sun's energy.

Non-renewable energy resources

Fossil fuels

Only a very small amount of the Sun's energy is trapped by plants and used to make food. The food is used by plants and animals. When plants and animals die, their bodies decompose and the energy that they possessed passes into decomposing organisms, such as bacteria, and eventually passes out into the air and space as heat. In the past, there were regions in the world where this did not happen. The energy in the bodies of ancient trees and marine organisms was trapped underground in their bodies. These organisms did not decay and **fossil fuels** (coal, oil and gas) were formed.

Today, fossil fuels are burnt in the boilers of power stations to generate electricity. The stored chemical energy in the fuel is transferred during combustion through the water in the boiler. The energy now has the form of kinetic energy, and moves the particles in the water so fast that they form steam. The kinetic energy in the steam is transferred to turbines in the power station, then the kinetic energy of the turbine blade is transferred to a magnet surrounded by a coil of wire in the generator. As the magnet spins, its energy is transferred to the coil of wire in the form of electrical energy.

◀ **Figure 17.13** A coal-powered power station.

7 Use a world map in an atlas to locate your country, then find it on the map on this page.
   a Roughly how long ago did people reach where you live now?
   b How long did it take people to migrate from Africa to where you live now?
   c When people left what is now India, which present-day countries did they pass through to reach
      i New Zealand
      ii South America?

Science in context

Our impact on ecosystems by using resources

Scientists believe that the human species first developed in East Africa between 300 000 and 200 000 years ago, gradually spread north over the next 100 000 years and then moved into the lands to the east and west.

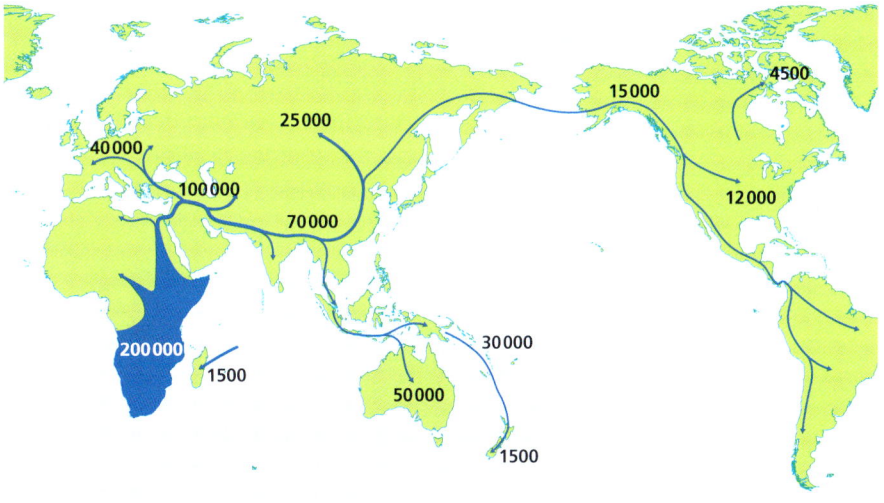

▲ **Figure 17.14** The arrows show the paths people took as they migrated across the world. The numbers show how long ago people reached the various parts of the world.

8 Explain why all the resources of hunter-gatherers were renewable.

9 Why did the presence of hunter-gathers in an eco-system have very little effect on it?

In Chapter 6 we looked at different ecosystems and studied the relationships between some of their different parts. In Chapter 7 we considered invasive species. Humans were certainly an invasive species into all the ecosystems of the world, but at first their effect was very small. They, like us, needed material and energy resources and, also like us, food resources. All their resources were renewable.

The first people lived as hunter-gatherers, roaming through their environment using the materials around them – like wood, stone and plant fibres – to make their homes and other items, which they used to collect food from plants and to hunt and catch animals. They used energy from burning wood to cook their food. People living as hunter-gatherers today are found in small numbers in the Amazon rainforest of South America, in some regions of southern Africa, as well as Thailand, Malaysia and Australia.

◀ **Figure 17.15** A temporary village of present-day hunter-gatherers in southern Africa.

About 10 000 years ago, people began discovering how to grow crops and how to farm animals and so they cleared their environments to make fields and farms. Farming provided food for larger numbers of people, who gathered in villages, towns and eventually cities. Farming also allowed people to work in other occupations, doing things such as pot-making, furniture-making, cloth-making and metal-working, and a trade of goods built up between neighbouring countries, reaching from Ireland in the west to China in the east. During this time, the sources of energy for farming, making goods and for transport were all provided by living things – the people themselves and the animals they had domesticated, such as cattle, horses and asses.

◀ **Figure 17.16** This view across the countryside in England shows how the forests have been cleared to make fields.

10 How do you think the Industrial Revolution affected ecosystems at the time?

Up until the late-eighteenth century, all goods were made by using simple machines, such as hand-operated looms for cloth or handsaws for cutting and shaping wood. At that time, it was discovered that steam could be used to power engines, called steam engines. These engines could give power to a large number of machines at once, so more goods could be made. This was the beginning of the **Industrial Revolution** and resulted in an ever-increasing use of resources of energy (at first coal) and of materials to make more goods to trade. The steam engines and machines operated in factories, which had chimneys that released large quantities of smoke and carbon dioxide into the air.

◀ **Figure 17.17** The smoky skyline of Glasgow in the mid-nineteenth century.

The Industrial Revolution spread across the world and, today, most countries have factories making products for their people and for trade.

The money earned from industry is used by countries to set up electrical supply networks, water supply networks and sewage systems. It is used by many people to live in homes in towns and cities. It allows many people to travel in cars and aeroplanes, and to shop for a huge range of goods in supermarkets.

◀ **Figure 17.18** Towns and cities around the world consume vast amounts of energy, food and materials to make all kinds of goods.

◀ **Figure 17.19** All of the engines in this photo use fuel derived from oil.

◀ **Figure 17.20** The goods on sale in supermarkets come from all around the world.

Assessing our impact on ecosystems

Human activity may influence ecosystems in a positive way, a negative way, or it may not influence them at all, in which case it is considered neutral. We can use these three ways of measuring our effect on ecosystems to give an activity an Ecosystem Impact Rating (EIR).

1 What is the EIR of hunter-gatherers – positive, negative or neutral? Explain your answer.
2 What is the EIR of early farmers – positive, negative or neutral? Explain your answer.
3 What is the EIR of setting up towns and cities – positive, negative or neutral? Explain your answer.
4 What is the EIR of the Industrial Revolution – positive, negative or neutral? Explain your answer.

Our lives depend on the use of these resources – food and water, energy and materials – and scientists and engineers around the world continually study them to see how they can be carefully used to maintain our way of life.

LET'S TALK

If you were to take up a career to try and use resources, but also reduce their Ecosystem Impact Rating, what would you do? Explain your answer.

CHALLENGE YOURSELF

Use the internet to find out about the major industries in your country and when the Industrial Revolution began there. If you have friends in other parts of world, ask them to do the same and compare how industry spread and developed in your countries. Make a report of your findings. Investigate several sources and look to see how they agree or disagree with each other. Bear in mind that some may be biased. If you feel they are, then mention this in your report.

Summary

- ✔ Renewable material resources can be replenished over time, but non-renewable materials resources cannot be replaced once used.
- ✔ Renewable energy resources include wind, tidal and solar power.
- ✔ Science in context: There are a variety of environmental problems involved in setting up a wind farm.
- ✔ Non-renewable energy resources include fossil fuels.
- ✔ Science in context: We can measure our impact on ecosystems by assigning an Ecosystem Impact Rating (EIR) to our activities.

End of chapter questions

1 **a** Explain why wind is a renewable resource.

   b How is energy taken from the wind and made useful to us?

2 **a** What are the tides?

   b Why is energy that is taken from the tides a renewable resource?

   c How is tidal energy transferred into a form of energy that we can use?

3 What is solar power?

4 **a** What renewable materials are used to make bioplastics?

   b Why should we be using more bioplastic and less ordinary plastic?

5 Why are fossil fuels a non-renewable resource?

6 How has the human Ecosystem Impact rating (EIR) changed since people moved from East Africa and spread around the world?

 Now you have completed Chapter 17, you may like to try the Chapter 17 online knowledge test if you are using the Boost eBook.

In this chapter you will learn:
- the difference between climate and weather
- that the Earth's climate can change due to atmospheric change
- about climates of the world (Science extra)
- about the cycle of the Earth's climate and the evidence of a cycle between warm periods and ice ages over long time periods.

Do you remember?

- What are the gases in the air?
- Describe the movement of particles in a gas.
- What do you understand by the word 'atmosphere'?
- Describe the weather today.
- Does the weather change during the year? Explain your answer.
- What do you understand by 'erosion in the rock cycle'?
- What is a fossil?
- Where are pollen grains made and why are they important?

CHALLENGE YOURSELF

Watch a weather report on your local television station. Note the features which are described in the report. Make observations on the weather for three days, then make a report on it. Your observations can range from just looking to using weather-recording equipment, such as a thermometer and a rain gauge. Did the weather change over the period of your observations?

Weather and climate

The **weather** is the description of the conditions in the atmosphere at a place over a short period of time, such as a day or a week. These conditions in the atmosphere include temperature, rainfall, sunshine, wind speeds and directions, and the amount of cloud. The **climate** is a description of conditions at a place over a much longer period of time, such as 30 years or more. Climate can be thought of as the average weather conditions in a place, usually quite a large area of the planet, over a long period of time.

Modelling the atmosphere

You will need:

an orange, or a ball of similar size, and a sheet of cling film (transparent plastic food wrap).

Process

1 Wrap the orange in one layer of cling film (transparent plastic food wrap).
2 The orange represents the Earth and the cling film (transparent plastic food wrap) represents the atmosphere.
3 What does this model tell you about the Earth and its atmosphere?
4 What are the limitations of the model?

1 An analogy is a comparison of one thing with another. In this model of the atmosphere, what is the analogy for
  a the Earth
  b the atmosphere?

▲ **Figure 18.1** The Earth from space.

The Earth's climate

The Earth's climate is due to its **atmosphere**. The features of the atmosphere are due to the heat energy transferred to it from the Sun. This warms the air and causes winds. The heat energy causes evaporation at the surfaces of the oceans, seas and lakes. This raises the amount of water vapour in the air which, when it cools, condenses to form rain.

The movements of the Earth affect the atmosphere. As the Earth rotates on its axis, it causes the air to move too. As the Earth moves in its orbit around the Sun, tilting sometimes towards the Sun and sometimes away from it, this causes the amount of heat at different places on the Earth's surface to change, which also causes movements of the air.

If you look at the Earth from space, you can see clouds moving in swirls over the planet, which indicates that the atmosphere is constantly changing.

The Earth's climate is sometimes called the **global climate**. Scientists use many different measurements to investigate and monitor the Earth's climate, as Figure 18.2 shows.

2 What would be the appearance of the Earth if the atmosphere did not change?

3 Where would the Earth have to be for the atmosphere not to change?

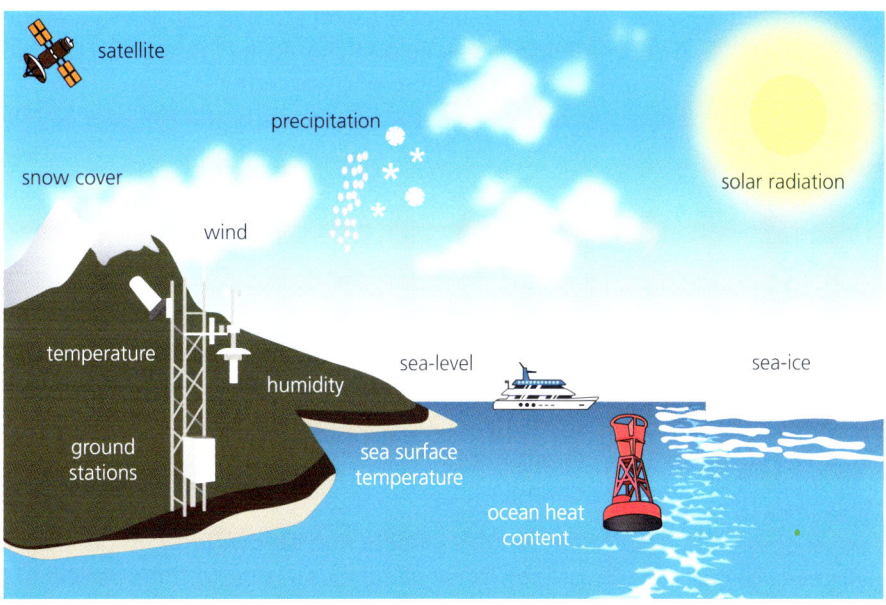

▲ **Figure 18.2** Collecting data on the Earth's climate.

Scientists use the data collected and apply it to a world map. Figure 18.3 shows the temperature changes in all regions of the planet from the data collected between 1988 and 2017. A colour code has been used to identify the change in temperature in a particular place.

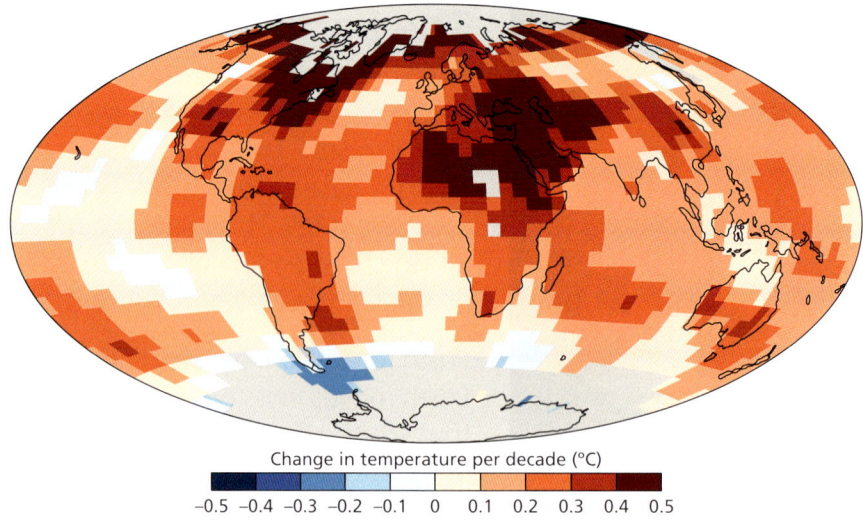

Change in temperature per decade (°C)

−0.5 −0.4 −0.3 −0.2 −0.1 0 0.1 0.2 0.3 0.4 0.5

▲ **Figure 18.3** Global temperature trends from 1988–2017. Source: NOAA Climate.gov.

The average (mean) yearly global temperature for the whole of the twentieth century was 12.7 °C. At the beginning of the third decade of the twenty-first century, it was 1.16 °C higher.

4   Use a map of the world to find out where you live on the planet. Now find the piece on the map in Figure 18.3. How has the temperature been changing there each decade?

5   Which colour is used to show the greatest increase in temperature?

Science in context

Climates of the world

From data collection studies around the world, like those shown in Figure 18.2, scientists have divided the Earth's climate into different climates across the globe. By doing this they can see over time how these climates change, which in turn gives an indication of how the Earth's climate is changing.

The Earth's climate can be divided into different climates across the globe. These climates are due to the position of the place on the planet, the structure of the land there and the movement of the atmosphere over it, such as the wind, and the clouds and rain they can bring. The map in Figure 18.4 shows the eleven major climates.

Here are brief details of the climates shown on the map.

Rainforest climate: hot and humid with rainfall almost every day.

Savanna climate: a warm climate which features two seasons – a wet season of heavy rain and a dry season with no rain.

CHALLENGE YOURSELF
Use the internet or news sources to find the average (mean) yearly global temperature this year. How does it compare with the temperatures
a   during the twentieth century
b   at the beginning of the third decade of the twenty-first century?

6 The major climates can be divided into subgroups. These have a modification to the basic information shown for each climate on the map in Figure 18.4. Look at a map of the world and find your country, then find its position on the map in Figure 18.4. Identify your climate, then research your specific climate details online. See if your research varies from the basic climate details given in this book. Look at a few sites that describe your climate. Do they all give the same information? For example, does a site aimed at tourists to your country present exactly the same information as a government site?

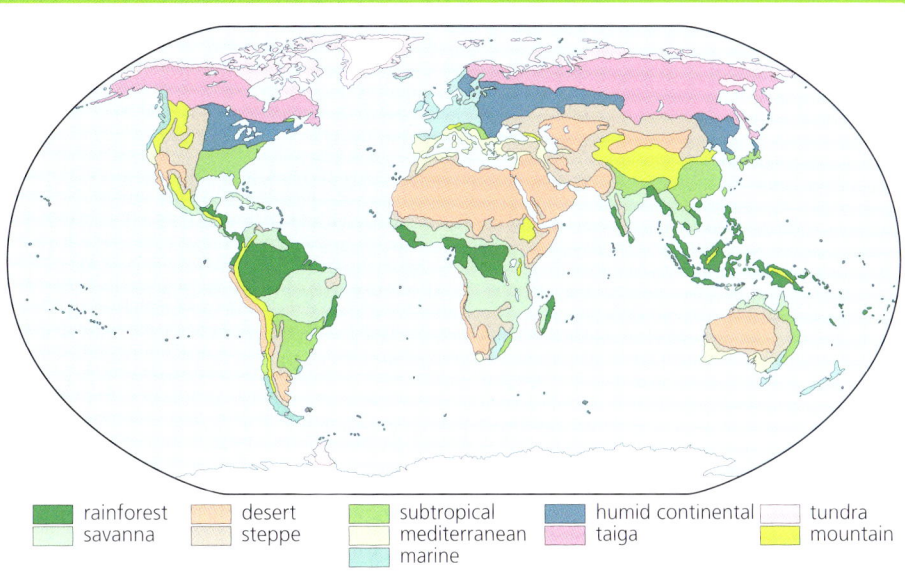

rainforest	desert	subtropical	humid continental	tundra
savanna	steppe	mediterranean	taiga	mountain
		marine		

▲ **Figure 18.4** A map of the Earth's climates.

Desert climate: long periods of dry weather with occasional short periods of rain, some desert climates are very hot in the day but other deserts are cool.

Steppe climate: warm or hot summers and cold or very cold winters with more rain than a desert climate.

Subtropical climate: hot summers which may be humid and warm or cool winters. There may be wet seasons and dry seasons. One type of subtropical climate is called the monsoon climate which has a long period of heavy rain.

Mediterranean climate: dry summers which may be warm or hot and cooler winters with some rainfall.

Marine climate: warm summers and cool winters with frequent changes in the weather with periods of rainfall throughout the year. Land with a marine climate is affected by the air which blows in from the surrounding sea. For example, the author lives in Great Britain which is shown in pale blue near the top centre of the map.

Humid continental climate: four very clear seasons – very cold winters, warm wet springtimes, warm humid summers, cool dry autumns (this season is sometimes called the fall).

Taiga climate: very cold winters up to six months long, short, warm or hot summers with rain.

Tundra climate: very cold dry winters with snow, cool summers with rain.

Mountain climate: cold all year with weather that changes quickly and features wind, rain and snow.

Climate change in the past

Scientists have discovered that the climate has changed many times in the past. These changes follow a pattern which can be described as a cycle of warm periods and cold periods. The cold periods are known as **ice ages**. Ice ages are divided into glacial and interglacial periods. We are currently living in an interglacial period, but we are also living in an ice age. Historically, warmer periods come in between ice ages, but we are not living in one of those. Rather, we are living in a warmer period within an ice age. The evidence for climate change has been provided by a range of studies, which include the study of rocks, fossils, pollen grains and ice cores.

Sedimentary rocks

▲ **Figure 18.5** Layers of sandstone some only a few centimetres thick which settled down almost 300 million years ago.

When rocks are exposed to the atmosphere, erosion takes place. The rocks break up into small fragments and are carried away by wind and water. In time, they form a layer known as a **sediment** and form sedimentary rock. Scientists study layers in sedimentary rocks because studies of the fragments can reveal what the climate was like when the erosion took place. The different sizes of fragments in different layers can indicate if there has been a change in sea level. A change in sea level is an indication of climate change as we know it today. Rocks can also provide evidence of climates in other ways. For example, the layers of sandstone (see Figure 18.5) which were laid down in the Permian period (298.9–251.9 million years ago) contain a large amount of iron oxide, which makes them red and indicates that the climate at the time was hot.

Fossils

Fossils are the remains of plants and animals which are preserved in rock from the time when they lived. Plants and animals need a set of environmental conditions. If the climate no longer meets the requirements of a plant or animal, it becomes extinct and the fossils that form in sedimentary rock tell us when the climate changed.

Coral is formed by tiny animals in the jellyfish group that make rocky homes in which they live. There are many types of coral, and many other organisms live on coral reefs with them. Coral is very sensitive to the temperature in its environment, so when the climate changes from cool to warm, or warm to cold, some corals and the organisms living with them become extinct, and their fossils leave a record of the changes, as shown in Figure 18.6.

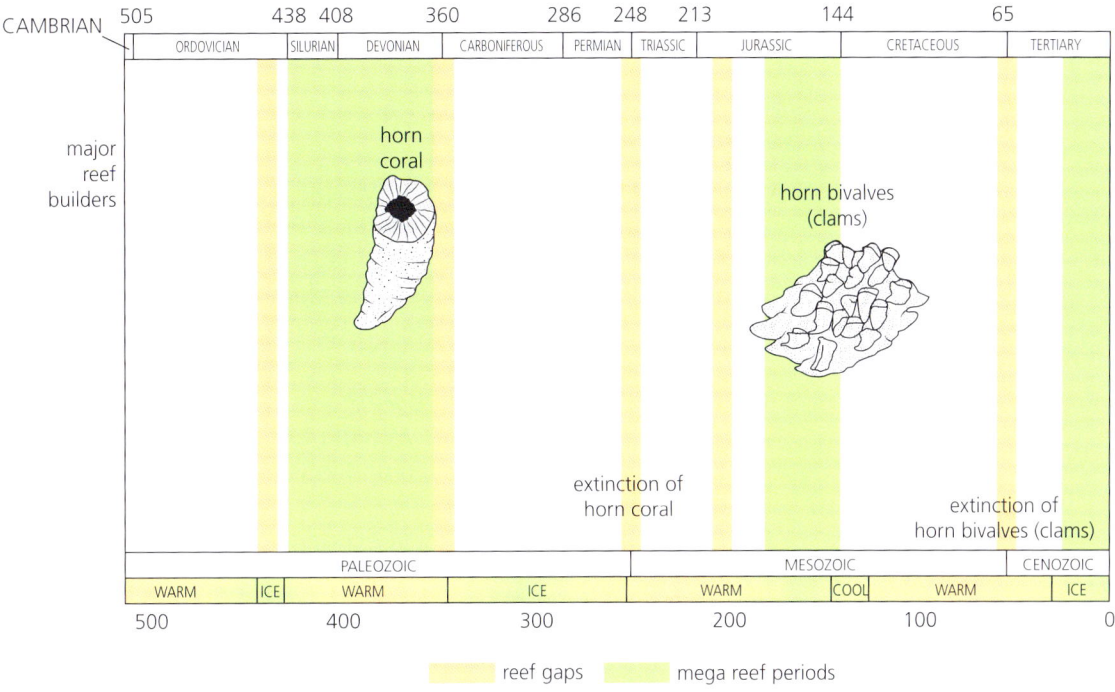

▲ **Figure 18.6** A fossil record of a coral and type of mollusc called a bivalve (clam). The numbers at the top show the ages in million years which you should read from right to left. The words below these show the names of the periods of time used by geologists. The three words below the fossils are larger periods of time used by geologists. The thick green bars show when the coral built a large number of reefs. The thin orange bars show when the coral did not build any reefs.

Pollen grains

Flowering plants and conifers produce pollen grains as part of their reproduction process. The coat of a pollen grain is made of a tough material that can be preserved in sedimentary rock or in ice.

The grains of different pollen found together as fossils indicates the types of plants living at that time, which in turn indicates the climate in which they lived. Different layers showing different combinations of pollen grains are used to describe the climate at the time when each layer formed.

▲ **Figure 18.7** This image shows a microscopic image of pollen grains. When fossilised pollen grains from tens of millions of years ago are found and studied under a microscope, it is possible to know which types of plants were alive at the time.

Can you make a pollen record for the future?

You will need:

paper, pencils, card glue, sticky tape, a microscope, a microscope slide, a pair of forceps (tweezers), a selection of local wild flowers that have been identified. (You may need to revise how to use a microscope and where to find the stamen and anther of a flower.)

Investigation and recording data

1 Pull a stamen off a flower with the forceps (tweezers).
2 Dab the anther on the slide and look for a yellow powder settling there.
3 Put the slide under the microscope and use the ×100 lens to focus on the pollen grains.
4 Draw a few pollen grains and state the plant they came from.
5 Repeat steps 1–4 with all the different species of flowers in the collection.
6 Set out the drawings on card for display and later storage. They could be used by students in a few years' time when they try this activity, to see if the plants have changed in your region in a short space of time.

Warning

People who are allergic to pollen grains must not try this activity.

7 How does the collection of pollen give an indication of the climate in your area and the global climate?

Ice cores

Ice cores are made by drilling into the ice and drawing out a long piece of ice. In these cores, scientists may find pollen grains, which suggest plants growing in a particular climate, dust from dust storms, which occur regularly in particular climates, and ash from volcanic eruptions, which can modify a climate.

◄ **Figure 18.8** The side view of part of an ice core showing a band of volcanic ash which settled out of the atmosphere about 21000 years ago.

Modelling an ice core

You will need:

a camera, crushed ice, a tall plastic cup, gravel, dried lentils, and access to a freezer (for your teacher).

Process

1 Take some crushed ice and let it form a layer in the bottom of the cup.
2 Sprinkle in some gravel to form a thin layer over the ice.
3 Place another layer of crushed ice over the gravel.
4 Sprinkle in some lentils of one particular colour.
5 Place another layer of crushed ice over the lentils.
6 Carefully pour cold water into the cup until it nearly reaches the top.
7 Let your teacher freeze your model core sample.
8 When you receive your model core sample, make a short video explaining how you made it, what the gravel and lentils represent, and how it shows how an ice core can be used to find out about climates in the past.

8 Do one or both graphs in Figure 18.9 show a pattern? Explain your answer.

9 Can you see a trend in both graphs? Explain your answer.

10 Look at the right-hand side of both graphs in Figure 18.9. Could the Industrial Revolution be responsible for what you see? Explain your answer. (You may need to look back at the previous chapter.)

Ice cores also contain tiny bubbles of gas from the atmosphere in the past. The gases in bubbles can be identified and analysed, and the amount of carbon dioxide in the gases can be plotted on a graph with an x-axis representing 800 000 years. A similar graph can be constructed from data in the rocks, fossils and pollen grains to show how the temperature has changed over the same period, as shown in Figure 18.9 below.

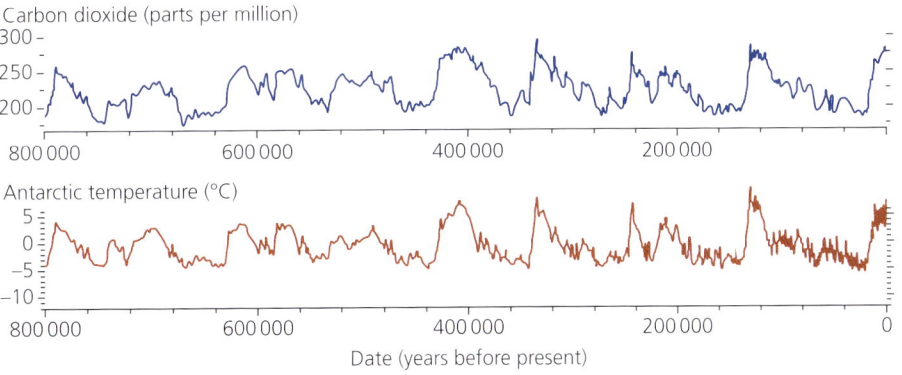

▲ **Figure 18.9**

Summary

✔ Weather is the description of the conditions in the atmosphere conditions at a particular place over a short period of time, such as a day or a week.
✔ Climate is a description of conditions at a place over a much longer period of time, such as 30 years or more.
✔ The Earth's climate can change due to atmospheric change.
✔ The Earth's climate follows a pattern which can be described as a cycle of warm periods and cold periods. The cold periods are known as ice ages. Ice ages are divided into glacial and interglacial periods.

CHALLENGE YOURSELF

What are tree rings and how can they be used to provide evidence of climate change? Use the internet or books to find out, and write a short report. Is their use limited? Explain your answer.

End of chapter questions

1 Name four features of the weather.
2 What does the word 'weather' mean?
3 What does the word 'climate' mean?
4 How does the Sun affect the atmosphere on Earth?
5 Name two ways the movement of the Earth makes the atmosphere change.
6 Name three things that scientists have studied for evidence of climate change in the past.
7 How has the Earth's climate changed in the past?

 Now you have completed Chapter 18, you may like to try the Chapter 18 online knowledge test if you are using the Boost eBook.

Inside a galaxy

In this chapter you will learn:

- about the Milky Way
- about galaxies and intergalactic space in terms of stellar dust and gas, stars and solar systems
- how to make model exoplanetary systems
- that asteroids are rocks that are smaller than planets
- about asteroids of the solar system (Science in context).

Do you remember?

- What is at the centre of the solar system?
- What else is in the solar system?
- How did a planet like the Earth form?
- What is the force that exists between the Sun and everything else in the solar system?

Stars

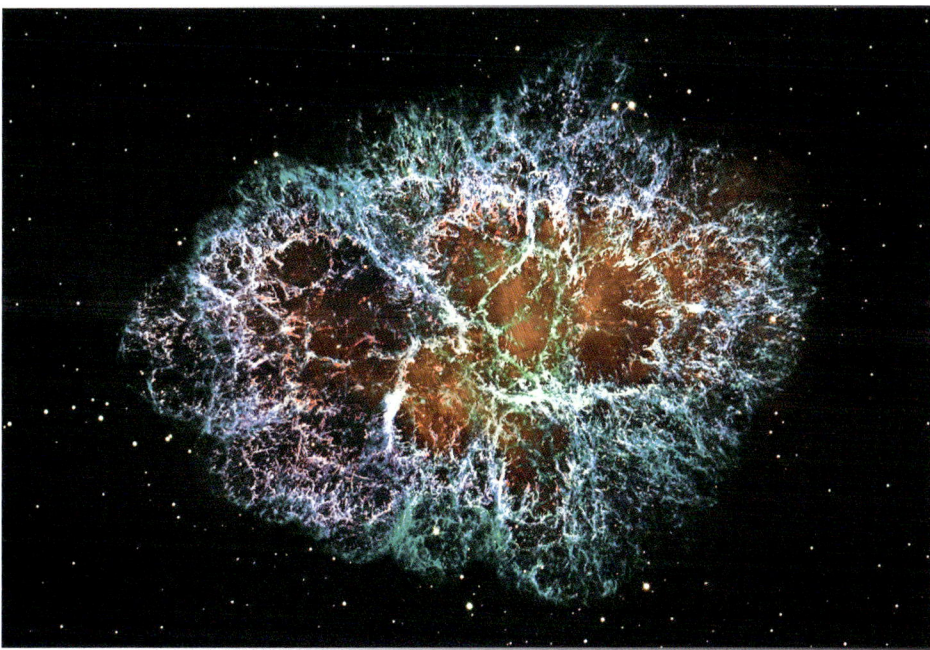

▲ **Figure 19.1** The remains of a supernova spreading out through space.

▲ **Figure 19.2** The Milky Way as seen from Earth.

The Milky Way

If you look up into the clear night sky, well away from street lights, you may see this view of space.

The huge white smudge running across the sky reminded early people of spilt milk, and so it became known as the Milky Way. In time, with the invention of telescopes, it was discovered that the Milky Way was composed of a huge number of stars. The astronomical term for a huge group of stars in space is a galaxy, of which the Milky Way is one.

Further studies on the Milky Way galaxy revealed that the stars are arranged in a spiral and that the Sun, our star, is in an arm of the galaxy, as Figure 19.3 shows.

a

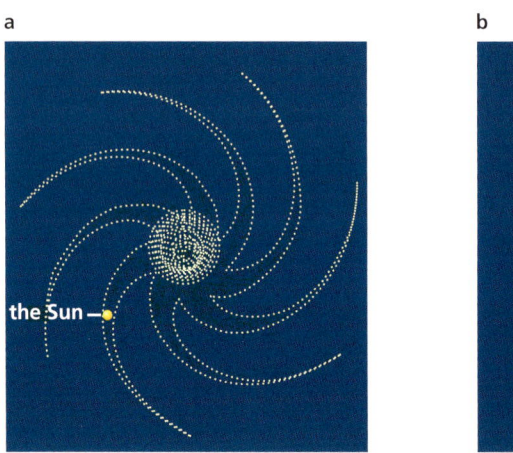

b

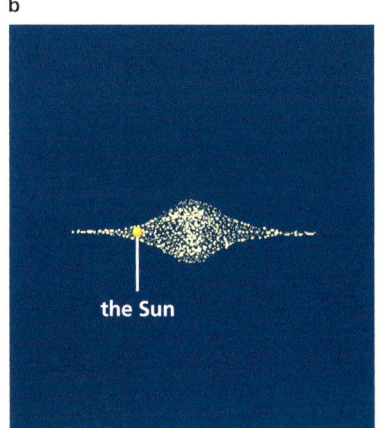

▲ **Figure 19.3** Side and top view of the Milky Way.

The Milky Way galaxy is about 100 000 **light-years** across and rotates like a huge pinwheel in space at 270 km/s. This means that the Sun, in its position about 25 light-years from the centre of the galaxy, travels at a speed of 828 000 km/h and takes between 225 and 250 million years to go round once. The last time the Sun was in its current position, dinosaurs were just starting to develop on the Earth!

Modelling a galaxy in a coffee cup

You will need:

a camera, a cup of cooled down coffee with no milk, two teaspoons, and a jug of cream.

1 Stir the coffee one direction in the cup.
2 When the coffee is swirling round, put half a teaspoon of cream on the second spoon.
3 Drop the cream off the teaspoon into the centre of the swirling coffee and record what happens on a camera.
4 What kind of galaxy do you make?
5 What was used for an analogy of the galaxy?
6 What is used for the analogy of space?

Galaxies

A galaxy is formed by stellar dust, stars and planetary systems that are all held together by the forces of gravity that exist between them.

Distant galaxies

As early telescopes improved, astronomers began to see smudges of light in the Milky Way galaxy. They discovered that they were clouds of gas in which stars formed. In the first half of the twentieth century, an American astronomer called Ernest Hubble (1889–1953) used the largest telescope then invented to discover that some of the gas clouds were in fact vast groups of stars far outside the Milky Way. These vast groups are separate galaxies, and today astronomers estimate that there are billions of them in the universe and that each one may contain over a billion stars.

▲ **Figure 19.4** Galaxies of stars as revealed by the Hubble space telescope, launched in 1990, which was named after the discoverer of galaxies beyond the Milky Way.

There are three types of galaxy (see Figure 19.5) – spiral galaxies, like the Milky Way and Andromeda, elliptical galaxies and irregular galaxies (which do not have a definite shape).

When astronomers discovered other galaxies in addition to the Milky Way galaxy, they also discovered that the galaxies were rushing away from each other, and that this helped to support the hypothesis of everything beginning with a huge explosion – the Big Bang.

▲ **Figure 19.5** Three types of galaxy.

Galaxies moving through space sometimes crash into each other. When this happens, the shapes of the galaxies change or they may merge to form a new one. If one galaxy is large and the other is small, the small one may be torn apart as the two collide. Small galaxies are thought to have crashed into the Milky Way galaxy in the past, and some scientists think there may be a chance that the Andromeda galaxy will crash into the Milky Way in a few billion years from now.

Science extra: Modelling the expanding universe

You will need:

the whole class, and each person will need a large black card to hold over their head, white paper, a pencil, scissors, a cushion, and access to an upstairs window to be used by a responsible person with a camera to film the 'universe' as it expands. If you have access to a drone, then it could be used instead of filming from an upstairs window.

Make this model in the following way:

1 Each class member should draw an outline of a galaxy shape on the white paper.
2 Each class member should then stick their galaxy onto their black paper.
3 Once the galaxies are complete, the members of the class should go out into an open area, preferably with a black tarmac surface.
4 The class members should come together into a close group and put the black paper above their heads, with the galaxies facing upwards.

5 The person filming from the upstairs window now gets into position and begins filming the cluster of galaxies.
6 On the count of three everyone should move outwards as if on the spoke of a wheel with the galaxy above their heads.
7 The film of the class' expanding galaxy should then be viewed and assessed as a representation of what scientists have observed.

Intergalactic space

The universe is composed of galaxies that are separated from each other by **intergalactic space**. In this space there is hardly any matter and scientists have estimated that one cubic meter (m^3) of intergalactic space has probably only one atom of hydrogen in it. This means that intergalactic space is almost a complete vacuum.

Stellar dust

If you were to travel through intergalactic space, you would find that as you approached a galaxy there is much more matter in the form of gases, such as hydrogen, and tiny solid particles called stellar dust. It is also called interstellar dust or cosmic dust. Stellar dust is produced by stars. It may range in size from just a few molecules joined together, to a rocky particle 0.1 μm across.

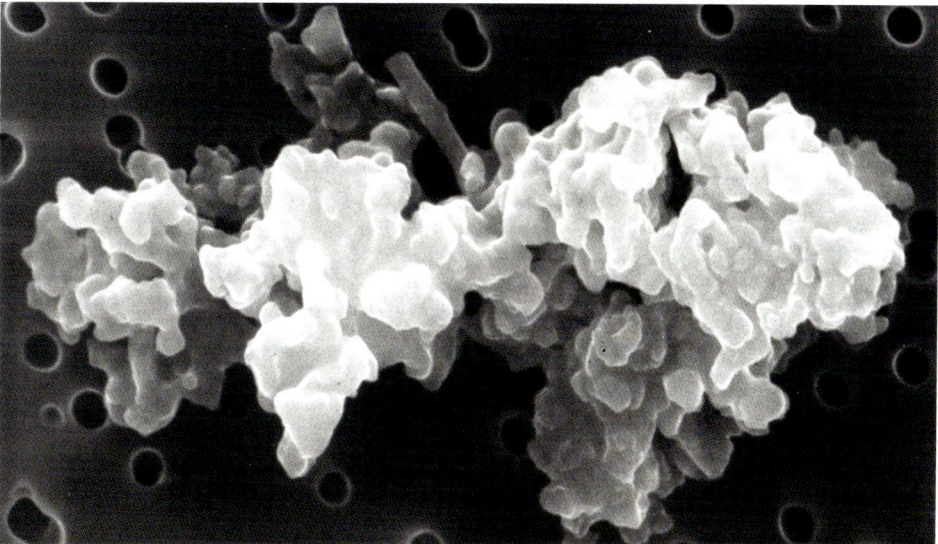

▲ **Figure 19.6** A particle of stellar dust as seen under a very powerful microscope. The particle in the middle is about two micrometres across (a micrometre is 1/1000000th of a metre).

Clouds in a galaxy

The gases and cosmic dust form huge clouds inside galaxies. Every galaxy has a number of gas clouds.

▲ **Figure 19.7** A gas cloud in a galaxy.

Planetary systems

Planets moving around a star are called planetary systems. Our sun and planets form a planetary system called the solar system. Sometimes the term solar system is also used for other planetary systems. A planetary system has a star at its centre, with one or more planets in orbit around it, some perhaps with moons, **dwarf planets**, asteroids and comets. The solar system is an example of a planetary system.

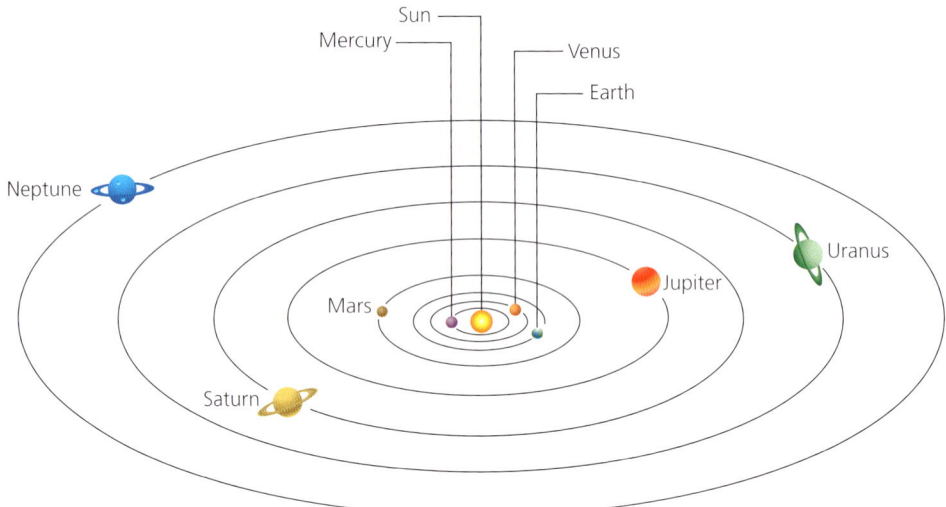

▲ **Figure 19.8** The solar system (not to scale).

The solar system was the only planetary system known until 1995, when a planet was discovered in orbit around another star. Planets in orbit around stars other than the Sun are called **exoplanets**.

Between 1995 and 2020, scientists discovered 3163 planetary systems. The number of systems with more than one exoplanet was 701, and the total number of exoplanets then discovered was 4281.

Figure 19.9 shows four planetary systems and their planets and orbits.

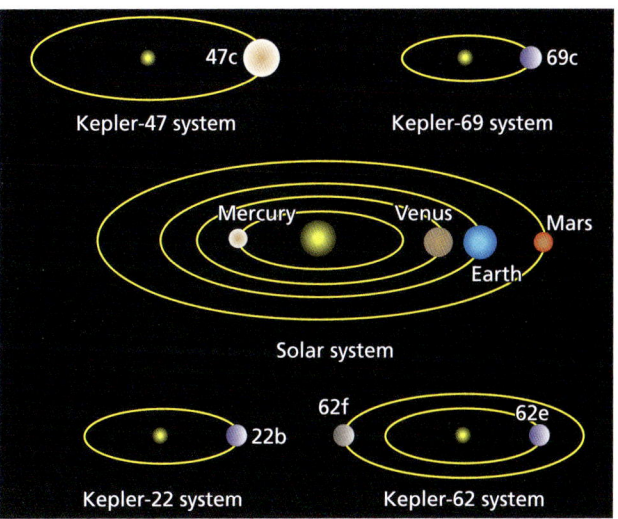

▲ **Figure 19.9** The four Kepler systems.

Modelling exoplanetary systems

You will need:

internet access, a camera, and a selection of fruit to represent a star and its exoplanets. For example, a melon could represent the star, and oranges, apples or other fruits of your choice could be used to make the exoplanets.

1 Begin by selecting two of the Kepler planetary system on this page, and make models of them with your fruit. Photograph each model before you use your fruit again for a different model.
2 Search the internet for pictures of other planetary systems. For example, TRAPPIST-1 in the constellation Aquarius has seven planets. Make models of each of the planetary systems you research and photograph them all.
3 For each planetary system you make, state the fruit you used as analogies for the star and its planet or planets.

Asteroids

When the solar system formed, not all the rocks that formed from cosmic dust went on to become parts of planets. They also formed smaller rocky objects called **asteroids** which move in orbits around a star, as planets do.

Asteroids of the solar system

Scientists have studied asteroids in the solar system and have produced this information about them.

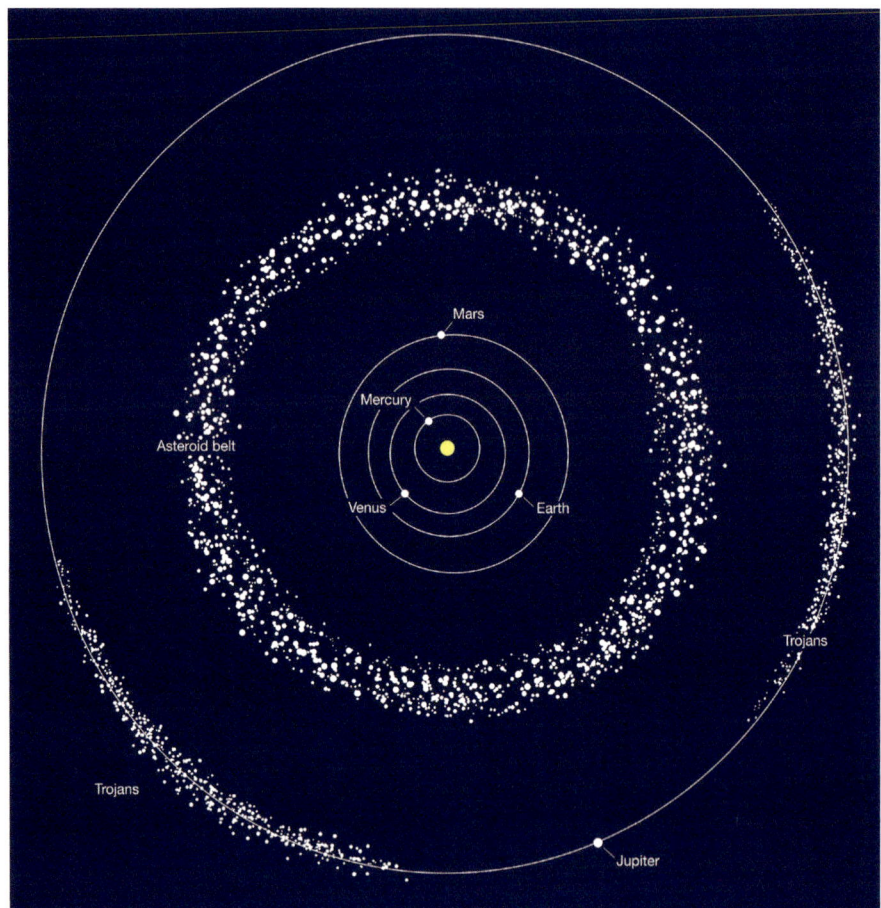

▲ **Figure 19.10** Asteroid belt and trojans.

Most of the asteroids have formed a circle, called the **asteroid belt**, between the orbits of Mars and Jupiter and they move in orbit around the Sun, while the others have formed two groups, called trojans, which are on either side of Jupiter and move in its orbit.

There are three groups of asteroid, classified according to the materials from which they are made. C-type asteroids (also known as chondrite asteroids) are the most common and have a dark appearance. It is believed that they are made from clay and rocks containing silica. S-type asteroids (or stony asteroids) also contain silica along with nickel and iron. M-type asteroids contain nickel and iron.

Asteroids range in size from as small as a grain of sand to Ceres, the largest asteroid, which is about 939 km across and is also classified as a dwarf planet. Not all asteroids are in the asteroid belt or in the trojan groups. Some have orbits further away from the Sun and a few have orbits that take them across the Earth's orbit.

Asteroids have their own gravitational field, as shown in Figure 19.11.

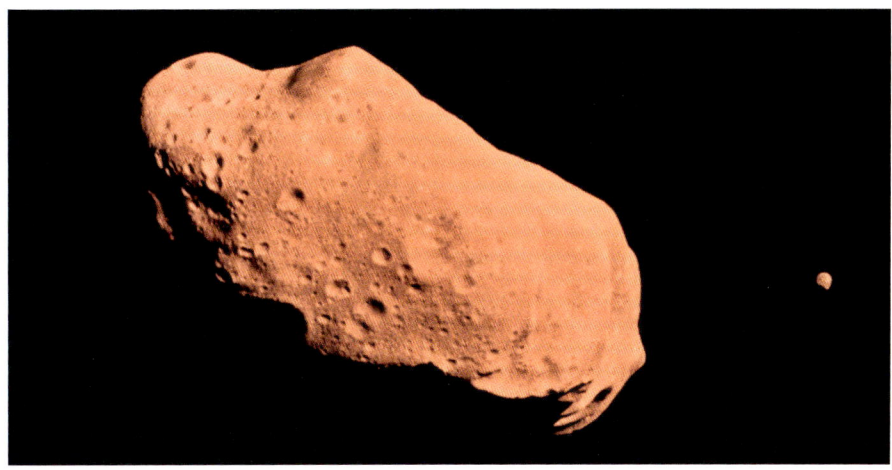

▲ **Figure 19.11** The asteroid Ida and its moon, Dactyl.

Summary

✔ The Milky Way is about 100 000 light years across and is composed of a huge number of stars.
✔ Modelling a galaxy in a coffee cup.
✔ Inventions like the Hubble Telescope allow us to view distant galaxies outside of the Milky Way.
✔ The movement of galaxies moving away from each other shows the expanding universe.
✔ Intergalactic space is almost a complete vacuum containing stellar dust as you approach galaxies.
✔ Planets moving around a star are called planetary systems, such as our solar system.
✔ Asteroids are formed from rocks left over from the formation of the solar system and they move in orbits around a star.

 Now you have completed Chapter 19, you may like to try the Chapter 19 online knowledge test if you are using the Boost eBook.

End of chapter questions

1 What is a galaxy?
2 What is stellar dust?
3 What is the diffence between a star and a planetary system?
4 Is the Earth an exoplanet? Explain your answer.
5 If you were investigating a planetary system, how would you recognise an asteroid?

 Glossary

A

aerobic respiration – a process which takes place in cells, in which energy is released from glucose when it takes part in a chemical reaction with oxygen.

allergic – Having an immune system which reacts to particular substances that are on pollen grains, animal hair, dust, or in certain foods such as nuts.

analogy – A comparison of one thing with something else.

antagonistic muscle pair – A pair of muscles in which, as one muscle contracts, causes the other muscle to relax or lengthen. For example, in the arm, when the triceps muscle relaxes, the biceps muscle contracts to make the arm lift.

antibody – A substance that attacks a pathogen.

artery – A blood vessel with elastic walls that carries blood away from the heart.

asthma – Condition which makes the tubes in the respiratory system narrower and increases the production of mucus, which makes breathing more difficult.

B

balanced diet – A diet in which all the nutrients are present in the correct amounts to keep the body healthy.

ball-and-socket joint – a joint between bones in which the end of one bone is round like a ball and fits into a hollow, called the socket, formed by the other bone.

blood vessels – Tubes in the bodies of animals that carry blood.

brain – The brain acts as the body's control centre. It is part of the nervous system and receives sense signals from nerves all around the body. It also sends information to the muscles and organs.

bronchiole – A small tube with muscles in its walls which connects the bronchus to tubes leading to the alveoli (the respiratory surface of the lung).

bronchus (bronchi) – Branch of the windpipe (trachea) that takes air into the lungs.

C

calorimeter – Equipment used to measure the amount of energy in a substance, by using the energy released when the substance is burned to heat water.

capillary – A blood vessel with walls that are one cell thick and through which substances (for example, oxygen and carbon dioxide) pass between the blood and the surrounding cells.

carbohydrate – A nutrient made from carbon, hydrogen and oxygen. Most are made by plants.

cardiac muscle – Muscle from which the heart is made that continues to contract and relax regularly, many times a minute throughout life.

cathode ray tube – A tube containing a vacuum through which a stream of electrons is allowed to pass to light up a screen coated with a light-sensitive material.

centrifuge – A piece of equipment in which test tubes of liquid mixtures are spun round at high speed to separate the different parts of the mixture.

chromatography – A process in which substances dissolved in a liquid are separated from each other by allowing the liquid to flow through porous paper.

combustion – A chemical reaction in which a substance combines with oxygen quickly and heat is given out in the process. If a flame is produced, burning is said to take place.

compound – A substance made from the atoms of two or more elements that have joined together by taking part in a chemical reaction.

creative thinking – Thinking freely about a question in science and producing lots of suggestions to test.

D

deficiency disease – A disease caused by the lack of a vitamin or a mineral in the diet.

diaphragm – A large sheet of mostly muscle which runs across the inside of the body just below the lungs. The diaphragm separates the heart and lungs inside the ribs from other organs that are lower down in the body. When the diaphragm contracts, it causes the lungs to expand and fill with air. When it relaxes, it causes the lungs to become smaller and push out some of the air they contain.

diffusion – A process in which the particles in two gases or two liquids, or the particles of a solute in a solvent, mix on their own without being stirred.

dispersion – The spreading out of light of different colours from a beam of white light, such as sunlight.

dissolve – The process in which a solute mixes with a solvent to form a solution.

distance/time graph – A graph showing the distance travelled by an object in a certain time. It can be used to find the speed of the object.

dwarf planet – A small planet with a small force of gravity that cannot remove small objects around it like larger planets can, but it may have moons around it.

E

electron – A tiny particle inside an atom which moves around the nucleus. It has a negative electric charge.

electrostatic – Describes the force which exists between two objects which have different electrical charges – one positive and the other negative, such as the protons (positive) and electrons (negative) in an atom.

element – A substance made of one type of atom. It cannot be split up by chemical reactions into simpler substances.

emphysema – A condition in which the gaseous exchange surface in the lungs has been reduced, resulting in breathing difficulties.

endothermic reaction – A chemical reaction in which heat energy is taken in.

energy – Something that exists in different forms (for example, electrical, light, sound and heat energy) that allows matter to move.

estuary – A large area of water at the mouth of river where it meets the sea or ocean.

exoplanet – A planet that moves around a star beyond our solar system.

exothermic reaction – A chemical reaction in which heat energy is released.

F

famine – A condition in an area where there is very little food, which results in starvation.

fats – Food substances that provide energy. They are made of carbon and hydrogen atoms linked into long chains with a few oxygen atoms and belong to a group of substances called lipids, which includes oils and waxes.

fibre – The undigestable part of food, called cellulose, which is found in plant cell walls and helps in the movement of food through the digestive system.

force – A push or pull that changes the way an object is moving, or makes it change shape.

fossil – The remains or impression of a plant or animal that lived in the distant past, preserved in rock.

fossil fuel – A fuel produced from the fossilised remains of plants and animals that lived long ago.

friction – The contact force that occurs between two objects when there is a push or a pull force on one of them that could make it move over the surface of the other. Friction acts to oppose movement.

fuel – A material such as coal, oil, gas or wood that is burned to release its chemical energy, producing heat and light. Alternatively, any source of chemical energy; for example, the body uses food as fuel to provide energy for life processes.

fulcrum – The supporting point around which a lever moves. It is also called the pivot.

G

gaseous exchange – The process by which oxygen moves from inhaled air into the blood and carbon dioxide moves from the blood to air that is about to be exhaled.

H

haemoglobin – The pigment in red blood cells that contains iron and transports oxygen around the body.

hinge joint – A joint between two bones which allows part of a limb to move in two directions, such as the lower arm being able to move up and down or the lower leg being able to move backwards and forwards.

hypothesis – An idea suggested to explain something. In science a hypothesis must be able to be tested by making a scientific enquiry.

I

Industrial Revolution – A period in history which saw a change from making things in homes and shops to making them in factories with the use of powered machinery.

inert – Being inactive; in chemistry, an inert substance is one that does not take part in chemical reactions.

inflamed/inflammation – a condition of part of the body, produced by the activity of white blood cells trying to heal tissues damaged by harmful microorgansims or chemicals, such as those in cigarette smoke, or any other foreign agent, such as a splinter.

inhaler – A device which is used to release chemicals into inhaled breath to widen the bronchioles and make breathing easier.

intergalactic space – The space in the universe between galaxies of stars.

L

light beam – A broad, straight path of light.

light-ray – A very narrow, straight path of light which in numbers forms a light beam.

light-year – The distance travelled by light across the universe in one year. This distance is 9.46 trillion kilometres, or 5.88 trillion miles.

luminous object – An object that releases energy in the form of light.

M

malnutrition – A condition of the body due to too much or too little of one or more nutrients.

marsupial – a type of mammal in which the final development of the offspring takes place in the mother's pouch.

mineral – A substance formed from an element or a group of elements. It may have a crystal structure.

mixture – An amount of matter made from two or more substances. Each substance is spread out through the other substance or substances.

model – Something which represents something else in the real world. A model helps us to understand something that is happening (in the case of a chemical equation), or the structure of something (in the case of a diagram such as that of a food web).

moment – A measure of the turning effect produced by a force around a fulcrum or pivot.

mucus – A substance produced by the lining of the respiratory system to trap dust and microorgansims.

N

neutron – A particle in the nucleus of an atom that has no electrical charge.

non-luminous object – An object that does not release energy in the form of light but may reflect light from luminous objects.

non-renewable energy resource – A certain amount of stored energy on the Earth that cannot be renewed when it has all been used up.

non-renewable material resource – A certain amount of material present on the Earth that cannot be renewed when it has all been used up.

nutrient – A substance in a food that provides a living thing with material for growth, development and good health.

O

obesity – A condition which develops in children and adults who have a high-energy diet but a lifestyle with low activity. This causes the body to store the excess energy as fat, which results in the body being greatly overweight.

organ – A part of the body, made from a group of cell tissues, that performs an important function in an organism.

organ system – A group of organs that work together to carry out a task to keep a living organism alive.

oxidation – The chemical reaction in which oxygen is added to a substance. The term is also used to describe the reaction in which hydrogen is removed from a substance.

oxyhaemoglobin – The substance formed when oxygen combines with haemoglobin. It transports oxygen around the body.

P

pathogen – An organsim that causes disease in other organsims.

philosopher – A person who applies critical thinking to any topic in order to further develop understanding or set out ideas to test.

pivot – The supporting point around which a lever moves, also called the fulcrum.

plane mirror – A mirror with a flat surface.

pollen grain – Microscopic grains produced by anthers, which contain the male gametes for sexual reproduction in flowering plants.

pressure – A measurement of a force that is acting over a certain area. It is measured in units such as newtons per metre squared (N/m^2).

prism – A transparent object which has at least two flat surfaces with an angle between them. Many prisms in science have three sides at angles and are called triangular prisms.

protein – A substance made from amino acids linked together in long chains. Proteins are used to build many structures in the bodies of living things.

proton – A particle in the nucleus of an atom that has a positive electrical charge.

pulse – The rhythmic expansion and contraction of blood vessels as blood is pumped through them by the heart.

R

reactivity – The speed with which a substance such as an element or compound takes part in a chemical reaction.

reflection – A process in which light-rays striking a surface are turned away from the surface.

refraction – The bending of a light-ray as it passes through one transparent substance into another.

renewable energy resource – A source of energy that can be replaced as it is used up, so that it does not run out; for example, energy from the Sun.

renewable material resource – A source of a material that can be replaced as it is used up, so that it does not run out; for example, wood.

research programme – A series of scientific activities, featuring experiments and enquiries, set up to investigate scientific topics and produce new scientific knowledge. Most programmes involve a number of scientists, often working in different parts of the world.

resource – A store of something which is useful, such as energy or a material such as wood.

respiration – The process occurring in all living organisms in which energy is released from food inside cells. Glucose reacts with oxygen to release energy for life processes, and carbon dioxide and water are produced. (Not be confused with breathing, which is the process of moving air in and out of the body.)

rusting – The formation of rust, which takes place when iron chemically reacts with oxygen and water.

S

scientific theory – An explanation of the things in our world such as phenomena, events, and the structure and working of living things that is based on a large amount of data (facts) and supported by repeat experiments and observations which leads scietists to believe that the explanation is true.

secondary colours (of light) – Magenta, cyan and yellow; created when light beams of the three primary colours overlap on a white screen. (White is created when all three primary coloured beams overlap together.)

skeletal muscle – Muscle which is attached to bones and moves them.

smooth muscle – Muscle which forms in the walls of tubes in the body, to control the size of the opening of the tube, and in the digestive system to move food along.

solubility – A measure of the ability of a solute to dissolve in a solvent.

solute – A substance that is dissolved in a liquid (solvent), forming a solution.

solution – A liquid (solvent) that has one or more substances (solutes) dissolved in it.

solvent – A liquid in which a solute can dissolve.

spectrum (visible) – The bands of coloured light seen when a prism disperses sunlight. The colours are red, orange, yellow, green, blue, indigo and violet.

speed – A measure of the distance covered by a moving object in a certain time.

starvation – A condition produced by lack of food where the body uses up all its own resources of energy and materials, which can eventually result in death.

steam – The gaseous state of water occurring at 100 °C.

subtropical – Areas just to the north and south of the Tropics of Cancer and Capricorn.

T

thought-showering – An idea-collecting exercise.

tide – The movement of the sea or ocean up and down the shore due to the pull of gravity of the Moon and the Sun.

tropical – Areas around the Equator and between the Tropics of Cancer and Capricorn.

tumour – An abnormal growth of cells that forms a lump which may or may not be cancerous.

turbine – A machine in which parts turn round and round and this movement is converted into electrical energy.

V

vaccine – A weakened or killed form of a microbe which causes a particular disease, which is injected into a organism to protect it from that same disease.

vacuum – A space in which there is no matter; it contains no atoms. Most vacuums are only partial vacuums in that they do contain a few atoms and molecules.

vein (animals) – A thin-walled blood vessel that transports blood towards the heart.

vitamin – A substance made by plants and animals that is an essential component of the diet to keep the body in good health.

vulnerable – in danger of being harmed. Populations of plants and animals that are vulnerable in their habitat are in danger of dying out and becoming extinct there.

W

weight – The downward force on an object due to gravity.

Acknowledgements

The Publishers would like to thank the following for permission to reproduce copyright material:

Photo credits

p.ix © Ryan McVay / Photodisc / Getty Images; **p.viii** *t* © CHARLES D. WINTERS/SCIENCE PHOTO LIBRARY; *b* © Pavel Losevsky / stock.adobe.com; **p.xii** *t* ©Andrew Lambert Photography/Science Photo Library; *b* © Aneduard/stock.adobe.com; **p.3** *l* © Anton/ stock.adobe.com; *r* © Thailoei92/Shutterstock; **p.4** *l* © Anton/stock.adobe.com; *r* © Eartty/stock.adobe.com; **p.8** © Светлана Фарафонова/stock.adobe.com; **p.9** © MICHAEL ABBEY/SCIENCE PHOTO LIBRARY; **p.13** *t* © Pressmaster/stock.adobe.com; *b* © Chanawit/stock.adobe.com; **p.18** © Moodboard/stock.adobe.com; **p.19** *t* © RFBSIP /stock.adobe.com; *b* © European Respiratory Society; **p.25** © SDI Productions/E+/Getty Images; **p.26** © Natika/stock.adobe.com; **p.29** *t* © New Africa/stock.adobe.com; *b* © BIOPHOTO ASSOCIATES/SCIENCE PHOTO LIBRARY; **p.37** © Chatuphot/shutterstock.com; **p.39** © Ton koene / Alamy Stock Photo; **p.40** © Farah Abdi Warsameh/AP/Shutterstock; **p.43** © Africa Studio/stock.adobe.com; **p.44** © Microgen/stock.adobe.com; **p.45** © The History Collection/Alamy Stock Photo; **p.46** *t* image credited to University of Minnesota School of Public Health; *b* Courtesy of University of Minnesota Archives, University of Minnesota - Twin Cities; **p.49** *t* © JackF / stock.adobe.com; *b* © BSIP SA/ RAGUET H. / Alamy Stock Photo; **p.51** © SB Arts Media / stock.adobe.com; **p.54** © Matthew J. Thomas / stock.adobe.com; **p.55** *t* © Vallerato / stock.adobe.com; *b* © Pictorial Press Ltd / Alamy Stock Photo; **p.56** *t* © Historic Images/Alamy Stock Photo; *b* Harry Leung, Research Assistant, Herpetology. Photo credit: Harry Leung; **p.57** © Hong Liu, Ph. D., Associate Professor; **p.59** *t* © Matyas Rehak/stock.adobe.com; *b* © Vaclav/stock.adobe.com; **p.60** © Stefan Scharf/stock.adobe.com; **p.61** © Sam D'Cruz/stock. adobe.com; **p.62** *t* © Choksawatdikorn/Shutterstock.com; *b* © Tom Goaz/stock.adobe.com; **p.64** © RHJ/stock.adobe.com; **p.66** © Pictureguy32 / stock.adobe.com; **p.67** *t* © Blickwinkel/Hartl/Alamy Stock Photo; *b* © John Scott/Alamy Stock Photo; **p.68** © Willyam/stock.adobe.com; **p.69** *t* © Dave Watts / Alamy Stock Photo; *b* © Stargrass/Shutterstock.com; **p.73** © Patrick Landmann/ Science Photo Library; **p.74** © Tui De Roy/Minden Pictures/Alamy Stock Photo; **p.75** © Ephotocorp / Sanjay Thakur / Alamy Stock Photo; **p.76** © Jacomina Wakeford/ICCE; **p.83** © Pavel Losevsky / stock.adobe.com; **p.88** © Photo Researchers / Science History Images / Alamy Stock Photo; **p.89** © North Wind Picture Archives / Alamy Stock Photo; **p.90** © Michael Gray/stock.adobe.com; **p.94** © Sarah Haigh; **p.98** © Diana Taliun/stock.adobe.com; **p.99** *all* © Andrew Lambert/ Science Photo Library; **p.103** © Rattiya Thongdumhyu/Shutterstock.com; **p.104** *t* © Mauro Fermariello/Science Photo Library; *b* © National Institutes of Health/Science Photo Library; **p.106** *l* © LAWRENCE MIGDALE/SCIENCE PHOTO LIBRARY; *r* © Curtis Kautzer/stock.adobe.com; **p.107** *both* © ANDREW LAMBERT PHOTOGRAPHY/SCIENCE PHOTO LIBRARY; **p.108** © ANDREW LAMBERT PHOTOGRAPHY/SCIENCE PHOTO LIBRARY; **p.109** © Russell Roberds / iStockphoto; **p.110** © MARTYN F. CHILLMAID / SCIENCE PHOTO LIBRARY; **p.115** *t* © Bilanol/ stock.adobe.com; *b* © Kumar Sriskandan / Alamy Stock Photo; **p.119** © Studiomode/Alamy Stock Photo; **p.120** © John Boud / Alamy Stock Photo; **p.121** © gasparij /123RF; **p.122** © Andrey Popov/stock.adobe.com; **p.124** *l* © Novak/stock.adobe.com; *r* © Mark_studio/stock.adobe.com; **p.125** © KEITH KENT/SCIENCE PHOTO LIBRARY; **p.129** *t* © Thirdkey/stock.adobe.com; *b* © Fotoflash/stock.adobe.com; **p.135** © Scott Barbour/Getty Images News/Getty Images; **p.137** © Michael Dwyer / Alamy Stock Photo; **p.140** © NASA/JPL; **p.141** © Nicky Rhodes/stock.adobe.com; **p.142** © Bettmann/Getty Images; **p.144** © Spyrakot/stock. adobe.com; **p.152** © MoiraM / Alamy Stock Photo; **p.154** © OlgaLIS/stock.adobe.com; **p.157** © Jjmillan/stock.adobe.com; **p.158** © VPales / Alamy Stock Photo; **p.162** *both* © ANDREW LAMBERT PHOTOGRAPHY/SCIENCE PHOTO LIBRARY; **p.163** © Theclarkester/ stock.adobe.com; **p.164** © Andriy Pogranichny/stock.adobe.com; **p.166** *all* © ANDREW LAMBERT PHOTOGRAPHY/SCIENCE PHOTO LIBRARY; **p.169** © GORDON GARRADD/SCIENCE PHOTO LIBRARY; **p.175** © Photodisc/ Photolibrary Group Ltd; **p.178** © Universal History Archive/UIG/Shutterstock; **p.181** © Apic / Hulton Archive / Getty Images; **p.182** © Pamela jungers/EyeEm/stock.adobe.com; **p.187** *l* © Jeffrey Coolidge / DigitalVision / Getty Images; *r* © ANDREW LAMBERT PHOTOGRAPHY/SCIENCE PHOTO LIBRARY; **p.189** © ANDREW LAMBERT PHOTOGRAPHY/SCIENCE PHOTO LIBRARY; **p.192** © ANDREW LAMBERT PHOTOGRAPHY/SCIENCE PHOTO LIBRARY; **p.195** © Charles Taylor / stock.adobe.com; **p.196** © CPA Media Pte Ltd/Pictures From History/Alamy Stock Photo; **p.197** *t* Science Museum/Science & Society Picture Library ; *b* Leonid_shtandel/stock.adobe.com; **p.199** © ALEX BARTEL/SCIENCE PHOTO LIBRARY; **p.204** © Kletr / stock.adobe.com; **p.205** © Ivan Kurmyshov/stock.adobe.com; **p.207** *t* © Matthew Oldfield Travel Photography / Alamy Stock Photo; *b* © Dinodia Photos/Alamy Stock Photo; **p.210** © IanChrisGraham / iStock / Getty Images Plus / Getty Images; **p.211** *t* © Dave stamboulis/Alamy Stock Photo; *b* © HENNING BAGGER / Staff / AFP / Getty images; **p.212** *t* © MARTIN BOND/SCIENCE PHOTO LIBRARY; *b* © NASA/SCIENCE PHOTO LIBRARY; **p.213** © Alexandre G. ROSA/Shutterstock.com; **p.214** *t* © Neil Harris / Alamy Stock Photo; *b* © Chris Lofty/stock.adobe.com; **p.215** *t* © INTERFOTO / History / Alamy Stock Photo; *c* © Anopdesignstock/iStock/Getty Images Plus/Getty Images; *b* © Dpa picture alliance archive / Alamy Stock Photo; **p.216** ©

Index